Palgrave Studies in Water Governance: Policy and Practice

Series Editors

Christian Bréthaut
Institute for Environmental Sciences
University of Geneva
Geneva, Switzerland

Thomas Bolognesi
GEDT
University of Geneva
Geneva, Switzerland

Looking at the issues of water governance through the perspective of the social sciences, books in the Palgrave Series in Water Governance take a global perspective on one of the key challenges facing society today: the sustainable development of water resources and services for all. In stepping away from the traditional focus on engineering and geophysics, the series takes a more holistic approach to both consolidate and generate knowledge that can be applied to different geographic areas by academics, researchers, policy-makers, NGOs and the private sector. This series emphasises the link between science and policy through considering water as a socio-ecological system, water and the territoriality of action, and water in the context of conflicts.

Yi hyun Kang

Climate Change Adaptation in River Management

A Comparative Study of Germany and South Korea

palgrave
macmillan

Yi hyun Kang
Université Saint-Louis-Bruxelles
Brussels, Belgium

ISSN 2947-910X ISSN 2947-9118 (electronic)
Palgrave Studies in Water Governance: Policy and Practice
ISBN 978-3-031-10485-5 ISBN 978-3-031-10486-2 (eBook)
https://doi.org/10.1007/978-3-031-10486-2

Cover credit: Daniel Bosma/Getty

This Palgrave Macmillan imprint is published by the registered company Springer Nature Switzerland AG.
The registered company address is: Gewerbestrasse 11, 6330 Cham, Switzerland

Foreword

In this book, Yi Hyun Kang tackles an issue of tremendous global significance: climate change adaptation strategies related to water management. Major flooding events are increasingly common and have come with a heavy price in terms of both lives lost and economic damage. The Intergovernmental Panel on Climate Change warns that as the climate further warms extreme weather events will become more common.

Kang illustrates the importance of water management as an aspect of climate change adaptation looking at historical and institutional developments internationally and nationally. She situates her work in the field of historical institutionalism, examining the power and influence of different actors and the water management policy solutions they pursue. Thus, while traditional planners and industrial interests argued for investments into grey infrastructure (e.g., the construction of dams and the laying of concrete walls along rivers), environmentalists have tended to give preference to green and blue infrastructure (e.g., wetlands and floodplains).

Kang examines Germany's and South Korea's national climate adaptation plans and flood prevention systems. In terms of South Korea, she argues that two discourses have dominated: developmentalism and techno-centrism although some shift towards greater environmental awareness is also recognizable. To examine these claims, she looks at South Korea's 2009 Green New Deal, which was heralded as a policy that would help make South Korea a front runner in the global green economy. A major component of the Green New Deal was the Four Rivers Project, a massive engineering project to construct 16 dams along the country's four major rivers (Han, Nakdong, Geum, and Yeongsan). The project followed a techno-centric approach to water management, meaning it aimed to control flooding through re-engineering of river flows with dams, weirs, and irrigation reservoirs. While the project also led to some ecosystem restoration and the addition of bike lanes, sports facilities, and other regional development initiatives along the rivers, it was criticized by environmentalists as very damaging to natural ecological systems. Their efforts to promote more integrated water management solutions and ecosystem protection as well as to incorporate more differentiated environmental justice perspectives were largely ignored by governmental planners. Environmental movements, she contends, remain relatively weak in Korea. Korea remains a developmental state, with a strong presidency and where the construction industry has a very strong influence; in contrast, the environment ministry and non-governmental organizations are quite weak.

In Germany, like in South Korea, there is a tendency towards technological optimism in planning processes, and this contributed historically to the straightening of rivers, like the Rhine, the building of canals and dams, and the removal of flood plains. She suggests, however, that in the past several decades, there has been a much stronger embrace of multi-stakeholder decision-making approaches and efforts to re-naturalize river flows and use agricultural and natural areas as floodplains. This shift she argues was possible because of a change in discourses tied to the relatively strong roles played by environmental actors as well as European Union influences. German water management is shaped by the European Union's Water Framework Directive and Floods Directive along with the national Water Management Act and the National Flood Protection Programme. After repeated major water catastrophes, planners began to pay greater attention to these discourses and integrated water management concepts have become more accepted and common. Devastating floods in 2013 along the Elbe River were a turning point in the rethink on flood management towards far greater emphasis on re-naturalization (giving space to the rivers).

This book contributes significantly to the literature on comparative climate adaptation and shows how and why understandings of best approaches to water management are changing. It will be of interest to those concerned with the effects of climate change, evolving water management policies, and the ways in which water management is being viewed in major economies in Europe and in Asia. The book is an important and fascinating read.

Professor of Environment and Climate Politics Miranda A. Schreurs
Technical University of Munich
Munich, Germany

Acknowledgements

First and foremost, I wish to express my sincere appreciation to my PhD supervisor, Miranda Schreurs. She has encouraged me to continue my research, sought the best ways to support me, and given me valuable advice on my study whenever asked. I also wish to show my gratitude to my second supervisor, Stefan Wurster, who has generously advised me on my dissertation. I am thankful to Jesko Hirschfeld, who kindly agreed to be my mentor and provided constructive advice for my study. I wish to thank Christian Djeffal for accepting to act as chair of the examining committee.

This book could not have been written without all the interviewees who accepted my interview requests and shared their precious time for the interviews. Thank you very much.

The financial contribution for the research activities is truly appreciated. The TUM Graduate School granted me a TUM-GS Internationalization Support fellowship for my research visit to Seoul National University and participation in the Third European Climate Change Adaptation Conference in Glasgow. I gratefully acknowledge the financial support through the German Academic Exchange Service's STIBET programme and Kongressreisenprogramm. My participation to the European Society for Ecological Economics 2015 was supported by the University of Leeds.

I would like to recognise the invaluable assistance from my colleagues and friends. I wholeheartedly thank Dongping Wang, Andrea Bues, Gloria Amoruso, and Andrzej Ceglarz for reviewing my chapters and giving moral support. Sehee Jung, Milan Chen, Natasha Donevska, and other colleagues have provided helpful comments on my research during the PhD colloquiums. My gratitude also goes to Jessica, Giorgia, Nella, Renee, Marisol, and Manar who always share their passion and insight on climate change issues. I would like to thank Mipsie and Pheiffer for helping me in editing and proofreading the first draft of the manuscript.

I also thank my editors at Palgrave Macmillan for their patient support and two anonymous reviewers who provided valuable comments to improve this book.

Last but not least, I wish to sincerely thank Andreas and my family for their support and love. They have patiently and constantly supported me during this odyssey. I cannot miss thanking Dicker and Oni for their companionship when I was writing at my desk.

Praise for *Climate Change Adaptation in River Management*

"Why do countries take different technical approaches to flood control? In this ground-breaking qualitative study of the cases of South Korea and Germany, Yi Hyun Kang uses discursive institutionalism to explore the key role of social discourse and discursive conflicts in bringing about major shifts to long-standing patterns of flood risk management (Germany) or not (Korea). A must-read for those interested in the sources and dynamics of continuity and change in water management responses to climate change."

—Vivien A. Schmidt, *Professor of International Relations, Pardee School of Global Studies, Boston University, USA*

"This book provides a thorough comparative analysis of the political and institutional factors that shape the differences in climate change adaptation approaches in South Korea and Germany. The cases of this study show that social discourses, decision-making structures, informal norms and shocking events have influenced the flood risk management policies. Participatory processes were found to increase the chance of policy change. The book makes a case for a more stakeholder-inclusive climate adaptation governance, since participation can reduce the uncertainty related to implementing a new policy. The book is essential reading, not only for scholars interested in comparative analyses, but also for policymakers and politicians working on water and climate governance issues."

—Dries Hegger, *Assistant Professor of Regional Water and Climate Governance, Copernicus Institute of Sustainable Development, Utrecht University, the Netherlands*

Contents

List of Abbreviations and Acronyms

ACF	Advocacy coalition framework
AFK	*Ständige Ausschuss Anpassung an die Folgen des Klimawandels*
AR	Assessment Report
BMU	*Bundesministerium für Umwelt, Naturschutz und nukleare Sicherheit*/ Federal Ministry for the Environment, Nature Conservation and Nuclear Safety
BUND	German Federation for the Environment and Nature Conservation
COP	Conference of the Parties
DAS	German Strategy for Adaptation to Climate Change
EC	European Commission
EEA	European Environment Agency
EIU	Economist Intelligence Unit
EU	European Union
EGLV	*Emschergenossenschaft/Lippeverband*
FGG	*Flussgebietsgemeinschaft*
GCF	Global Climate Fund
GDP	Gross domestic product
GEF	Global Environment Facility
GHG	Greenhouse gases
IAD	Institutional Analysis and Development
ICPDR	International Commission for the Protection of the Danube River
ICPER	International Commission for the Protection of the Elbe River
ICPO	International Commission for the Protection of the Odra River Against Pollution

ICPR	International Commission for the Protection of the Rhine Against Pollution
IPCC	Intergovernmental Panel on Climate Change
IWG	Inter-ministerial working group
IWRM	Integrated Water Resource Management
KEI	Korea Environmental Institute
KFEM	Korean Federation for Environmental Movements
K-water	Korea Water Resources Corporation
LANA	National/Federal Consortium for Nature Protection, Landscape Conservation and Regeneration
LAWA	Working Group of the Federal States on Water Issues
MOLIT	Ministry of Land, Infrastructure and Transport
NAP	National adaptation plan
NAS	National Adaptation Strategy
NASA	National Aeronautics and Space Administration
NCCAP	National Climate Change Adaptation Policy
NFPP	National Flood Protection Programme
NGO	Non-governmental organisation
NOAA	National Oceanic and Atmospheric Administration
OECD	Organisation for Economic Co-operation and Development
POMAC	Professors' Organization for Movement against the Grand Korean Canal
REDD+	Reducing Emissions from Deforestation and Forest Degradation
SDGs	Sustainable Development Goals
SREX	Special Report on Managing the Risks of Extreme Events and Disasters to Advance Climate Change Adaptation
UBA	*Umweltbundesamt*
UK	United Kingdom
UN	United Nations
UNECE	United Nations Economic Commission for Europe
UNEP	United Nations Environment Programme
UNFCCC	United Nations Framework Convention on Climate Change
US	United States
WFD	Water Framework Directive
WHG	*Wasserhaushaltsgesetz*
WMO	World Meteorological Organization
WWF	World Wildlife Fund

List of Figures

List of Tables

1

The Political Dimension of Water Management in the Face of Climate Change

1.1 General Introduction

Adaptation[1] to the changing climate is critical to human society. Even if greenhouse gas concentrations stabilise in the near future, the ongoing climate change is unavoidable. The impact of anthropogenic climate change is observable almost everywhere. The summer season in many countries is hitting record temperatures, resulting in a number of wildfires and droughts (Vogel et al., 2019; Barriopedro et al., 2011; IPCC, 2021). In the meantime, destructive storms, typhoons, tropical cyclones, and hurricanes are being reported more frequently (O'Gorman & Schneider, 2009; Trenberth, 2011; IPCC, 2021). The need for appropriate adaptation measures is growing fast as climate change accelerates. Adaptation to climate change has become an important subject for international politics (Berrang-Ford et al., 2011; Grothmann & Patt, 2005).

Among various subjects related to climate change adaptation, water is a particularly interesting and important one. Climate change has even been described as 'water change' because water is the principal medium

[1] See the definition of adaptation in Sect. 3.1.2.

Y. h. Kang, *Climate Change Adaptation in River Management*, Palgrave Studies in Water Governance: Policy and Practice,
https://doi.org/10.1007/978-3-031-10486-2_1

through which climate change will affect society (Bullock et al., 2009; OECD, 2013). More extreme and frequent weather events such as torrential rain and storms are expected in many countries, increasing flood risks (Milly et al., 2008; Huntjens et al., 2011). At the same time, periods without precipitation and the severity and length of heatwaves and droughts are likely to be extended (Wilhite et al., 2014). This makes water management[2] one of the most relevant issues in the adaptation policy field, which includes both anticipatory and reactive actions for reducing risks and damages from climate change (see Sect. 3.1.3 for more details of adaptation policy) (Pahl-Wostl, 2015; Bauer & Steurer, 2015).

Two interesting countries to study in relation to adaptation in water management are South Korea (hereafter Korea) and Germany, because of their contrasting policies despite having similar geographical conditions. These two industrialised countries are located in mid-latitude regions known for their temperate climate and are expected to experience more intense precipitation and prolonged dry periods as climate change proceeds (Korea Meteorological Administration, 2014; Deutschländer & Mächel, 2017). Due to high levels of industrialisation and population density, flood control and drought prevention are a high priority in both countries. Another similarity shared by the countries is that large river basins are the primary component of the landscapes. South Korea has four major rivers, namely the Han, Nakdong, Geum, and Yeongsan, while Germany has five major rivers including the Danube, Rhine, Elbe, Weser, and Oder. Although the two countries have different flood seasonality, they both face higher fluvial flood risks because of climate change. When the focus is placed on flood risk management in rivers, however, a clear difference exists between South Korea and Germany.

In South Korea, the 'Four Rivers Project' was implemented as the main adaptation scheme between 2009 and 2012. This project consisted of dredging riverbeds and constructing 16 dams along the country's four major rivers: the Han, Nakdong, Geum, and Yeongsan. This engineering-oriented project aimed to control the flood and low water level situation.

[2] Water management in this book refers to different political and technical approaches to the control and use of water resources. It is used as a broader term than water governance, which emphasises the political and social aspects of water management.

It was also a part of the Green New Deal policy introduced by the government of Lee Myung-bak[3] that at the time of its introduction was touted as a project that would create 340,000 jobs in the middle of the financial crisis (Ministry of Land, Transport and Maritime Affairs, 2009). The Four Rivers Project was in line with Korea's long-term river management policy since the 1960s, which has strongly focused on grey infrastructure and engineering methods.

In contrast, the German government has turned its direction from its long-standing technological approach to more nature-based solutions for flood control. Germany heavily modified its rivers in the past, just like South Korea. About two-thirds of former flood plains in Germany were reclaimed and many rivers were canalised (Blackbourn, 2006; Federal Government, 2015). This was done in order to control floods, utilise the land for farming, build industrial facilities, and create waterways. Now the German government is instead emphasising the re-naturalisation of its rivers. 'Preventive Flood Protection' within the National Flood Protection Programme (NFPP) is a special framework for expanding floodplains and giving more space to the rivers (BMU, 2017).

Germany and Korea share similarities in terms of not only geography and water management history, but also economy and politics, making the difference between the two countries' flood risk management more intriguing. Both countries went through fast development periods under strong economic growth imperative and achieved high level of industrialisation after the Second World War. Environmental quality and conditions deteriorated in both countries during the fast industrialisation period. The environment is now better protected in those countries, but institutional legacy and ideas have a great influence over environmental policies in both countries. In regard to politics, Germany and Korea have relatively high levels of democracy which is measured by levels of political participation, free and fair electoral processes, and civil liberties (EIU, 2018) (see Sect. 1.5.3 for more details).

[3] In this book, Korean names are written with the surname first as in Korean. In addition, Korean authors' names are written in full names with the surname first when it is necessary for better distinction.

The research question that guided this book raised at this point: *Why have South Korea and Germany pursued different adaptation policies for and approaches to water management, while they have similar geographical, economic, and political conditions?*

To answer this question, this book investigates the policy processes behind the water management policies stated in the initial climate change adaptation strategies of the two countries formulated in a similar period (the late 2000s).[4] Designed on a grounded theory approach (see Sect. 1.5.1), this research used hypotheses as a means of guiding the research direction, not as a supposition to test. The main hypothesis of this investigation is that discourses and institutions are the two main elements of policy process. Therefore, the analytical framework of the investigation, which will guide the analysis of empirical findings in a systematic way, is mainly based on discursive institutionalism (see Sect. 2.1.3 for the definition). Data collection and analysis were conducted following a qualitative approach. In the course of answering the research question, this book seeks to establish a theory regarding the decision-making process of adaptation politics beyond the two specific cases of South Korea and Germany.

1.2 Political Analysis of the Adaptation Policy Process

Climate change adaptation is a growing field of study, as the impact of climate change is becoming more intense and frequent. Interest in adapting to climate change has increased since the beginning of the twenty-first century. Although the term 'adaptation' was used in the Rio Declaration agreed in the United Nations Conference on Environment and Development (Rio Summit) in 1992, at the time the importance of adaptation was not widely recognised either by practitioners or by researchers (Pelling, 2011). As climate change impacts become more visible around the world and it has become clearer that climate change impacts cannot be completely avoided, interest in adaptation has grown

[4] The first German Adaptation Strategy was set up in 2008 and the first National Climate Change Adaptation Plan of South Korea was established in 2010.

(Fünfgeld & McEvoy, 2014; Moser & Ekstrom, 2010). Starting in the mid-2000s, many national governments began to develop national climate change adaptation plans (UNFCCC, 2018). The number of scientific articles in the adaptation field more than doubled in the period between 2005 and 2010 (IPCC, 2014).

Although all adaptation policies share the same goal of mainstreaming climate change adaptation (Bauer & Steurer, 2015), the forms and contents of the policies vary. However, there are still few analyses examining why countries choose certain adaptation policies, either in general or in specific sectors. While the scientific uncertainty associated with climate change projections has been widely discussed (Smith & Stern, 2011; Hallegatte, 2009; Wilby & Dessai, 2010), the uncertainty within the decision-making process on adaptation issues has not gained much attention in the adaptation field. Instead, adaptation responses have been treated as inevitable outcomes of scientific findings, rather than as political constructions (Eriksen et al., 2015; Nightingale, 2017). This might be attributed to the dominance of science and engineering-based approaches in the adaptation field.

Adaptation has developed slowly into a policy agenda which includes socio-political aspects. Research on adaptation has increasingly addressed how approaches to development and political-economic systems may obstruct transformative adaptation (Pelling, 2011; Scoones et al., 2015; Bosomworth, 2018). Nevertheless, a reluctance to discuss the politics of adaptation exists in the field. There is still a clear gap in the literature regarding how informal institutions and power structures influence decision-making processes related to adaptation policies (Eriksen et al., 2015; Eakin & Lemos, 2006).

This book tries to show how governments develop adaptation policies based on a discursive institutionalist framework. Although climate change adaptation is a relatively new field, many phenomena that climate change exacerbates such as floods, droughts, and heatwaves have always existed. Many of the policies that are now part of climate change adaptation strategies are not entirely novel. Instead, the adaptation policies can be regarded as the product of policy development in each sector over time (Wise et al., 2014). Therefore, it is necessary to examine sectoral policy developments, in order to understand the contextual background of contemporary adaptation strategies and plans. However, not many studies on adaptation have dealt with older policies which were adopted and implemented before official adaptation plans (Füssel, 2007; Smit & Wandel, 2006; Dupuis & Biesbroek, 2013). In order to fill the gap, this book deals with each water management policy not just from the point at which adaptation policies were formulated, but over the last 60 years, approximately from 1960 to 2020. This time frame was decided based on the history of the modern water management policies in Germany and Korea. In Germany, the Federal Water Act (*Wasserhaushaltsgesetz*, WHG) was first legislated in 1957, whereas the 10-Year Plan for Comprehensive Water Resource Development in Korea was established in 1965.

Within political science, discussions on the politics of adaptation are scarce. Particularly, if we look into the field of studies on environmental politics within political science, adaptation has not been given much attention. Although several comparative studies on national adaptation policies exist (e.g., Bauer & Steurer, 2015; Lesnikowski et al., 2016; Greiving & Fleischhauer, 2012; Mertz et al., 2009), the number of such studies is still limited (Dupuis & Biesbroek, 2013). Ford et al. (2011) point out that the short history of adaptation policy and limited data are major reasons why cross-case studies and comparative studies for this subject are not abundant.

This book takes a discursive institutionalist approach in analysing the adaptation policies of South Korea and Germany. Social constructivist and discursive institutionalist approaches to adaptation policy analysis are relatively new in the adaptation field. A few examples exist (Lindseth, 2005; Risbey, 2008; Juhola et al., 2011), but the number of analyses on discourses and framing in the climate change adaptation field is underwhelming (Juhola et al., 2011). The reason why this study took discursive institutionalism as its analytical framework is explained in detail in Chap. 2.

1.3 Water: A Critical Factor in Climate Change Adaptation

The impact of climate change can be directly felt by changes in water. The intensification of the water cycle due to a warmer climate may result in more frequent and intense extreme weather events, increasing flood risks. Concurrently, droughts are likely to occur more frequently in presently dry regions, and freshwater provision could be interrupted (IPCC, 2014, 2021). This is why water is often referred to as the medium through which climate change impacts are felt. The intensity of the impacts of climate change on human society will be closely related to how effectively water is managed in all of its many forms. Bauer and Steurer (2015) assert that water management is suited for adaptation policy analysis because water management is one of the most relevant subjects of adaptation.

Floods and droughts are intensifying as climate change affects both mean and extreme values of "precipitation, evapotranspiration and rates of discharge of rivers" (Milly et al., 2008, p. 573). Severe floods can lead to massive losses and damages to societies. In addition, climate change is making it harder to make accurate predictions about water resource availability. As predictions become more uncertain, conflicts surrounding water for different uses could increase (Pahl-Wostl, 2015).

Development of policies for adapting to changing water patterns has become a major concern in many regions. In the *White Paper—Adapting to Climate Change: Towards a European Framework for Action* published by the European Commission (EC) in 2009, for instance, water resource management was emphasised as the key strategy for enhancing resilience to climate change (European Commission, 2009). European countries put water as their top priority for adaptation (European Environment Agency, 2014).

Although water management is a popular issue in the adaptation field, the discussion has often been weighted towards establishing rules and developing and regulating infrastructure, with little attention given to the policy process (Mosello, 2015; Tanaka et al., 2006; Pittock, 2011). This is problematic because water management is political by nature and it is essential to understand the political dynamics behind certain policies.

For example, water management involves decisions related to the distribution of benefits and burdens: Who has access to fresh water? Which areas will be inundated by flood waters and which areas should be protected? These types of questions cannot be answered without considering the policy process, which is fundamentally political. Kaika (2003) points out that water management is one of the most critical political issues in Europe and the international community.

In the water management literature, the policy process and associated political conflicts have long been areas of interest (Pahl-Wostl et al., 2010; Moss & Newig, 2010). For example, discursive conflicts related to flood risk management measures are not new (Meijerink & Huitema, 2010). However, adaptation is a relatively new policy subject within water management. Although there have been research projects to investigate water policy in depth in order to design more sustainable flood risk governance (e.g., Alexander et al., 2016; Firus et al., 2011), there are few studies in the existing water management literature which deal with the climate change adaptation policy process and its relevance to water management. Therefore, the need for more integrative research which combines water management and adaptation policies through the lens of politics is growing.

In order to meet such increasing needs for integrative analysis, this book proposes a typology of water management discourses based on a literature review. They are (1) engineering-development, (2) participation-equity, (3) conservation-ecosystem service, and (4) integrated water resource management (IWRM). The engineering-development discourse has been the most common discourse in water management with the underlying confidence in controlling and predicting water systems (Trottier, 2004). Meanwhile, the conservation-ecosystem service discourse emphasises that restoring or keeping the natural state of water bodies provides significant benefits to the ecosystem as well as human societies. Environmentalists who hold this discourse argue that engineering-based solutions can be non-adaptive to the changing environment and even have negative impacts on social-ecological systems (Perini, 2017; Depietri & McPhearson, 2017; Huitema & Meijerink, 2009). The third discourse on water management, participation-equity, points out that relevant stakeholders should be engaged in decision-making processes in order to make water management policy fair and just. This discourse gained more support as environmental movements have grown in strength and the perception of governments' roles has changed within water management systems (Ingram, 2011; Pahl-Wostl, 2015). Lastly, the IWRM discourse promotes the balance of economic interest, social welfare, and ecosystem protection (Global Water Partnership, 2011). The first three discourses form a triangular discursive space in which various policy discourses can be located. For example, the IWRM is placed in the middle of the triangle. Section 3.2.3 elaborates the proposed water management discourse typology and the triangular discursive space in detail. This typology of water management discourses will be potentially useful for examining other cases beyond the two countries presented in this book.

1.4 Flood Control as a Continuous Battleground of Discourses

Among various sub-categories related to water management and climate change adaptation, this book focuses on flood risk management. This is first because both Germany and South Korea will face higher risks of flooding due to climate change, and their adaptation plans explicitly include a flood risk management policy. Second, research on the social or political aspects of flood risk management is still rare, while natural and technical science perspectives have dominated the field (Driessen et al., 2016; Dieperink et al., 2016). Therefore, there is a knowledge gap to which this research can contribute. Third, the investigation into the conflicts within the policy process of flood risk management in the last 60 years can provide insight into the policy process in general, which could be useful in the broader context of climate change adaptation.

Because there is no universal remedy or a 'one-size-fits-all' solution in water management (Hegger et al., 2016; Thieken et al., 2016), flood control is frequently a subject of dispute. Socio-political and economic factors are as important as physical and geographical factors in deciding flood risk policies, and such complexity makes it hard to reach a consensus on a specific measure (Driessen et al., 2016). Flood control is therefore still an ongoing battleground of politicians, stakeholders, and experts in many countries.

Methods of flood control in river basins have developed and diversified over time. In industrialised countries, structural measures based on technological and engineering methods dominated the field until the 1980s. Large-scale structural infrastructures had many advantages: canalised streams could prevent floods particularly in populated areas. Dams could be utilised not only for flood control, but also for power generation. Floodplains could be turned into industrial or agricultural areas, which contributed to increased productivity.

As environmentalism, which prioritises environmental protection, became a more powerful social discourse in industrialised countries in Western Europe and North America in the 1970s, concerns about water quality and ecosystem conservation grew sharply. Canals and dams had negative impacts on the ecosystems. Fish cannot migrate if the rivers are fragmented. Besides, engineers found that higher dikes may result in more severe floods as dikes can increase the flood peak. Furthermore, the erosive capacity is increased in straightened rivers, heightening flood risk in downstream regions (Plate, 2002). A 'lock-in' effect, which increasingly hinders changes from the status quo, is another problem of such large-scale, inflexible infrastructures in rapidly changing environments (Huitema & Meijerink, 2010).

Since the 1980s, therefore, approaches supporting controlled flooding through natural floodplains and retention areas began to attract the attention of policymakers, especially in Western Europe and the United States. Non-structural and green and blue infrastructure for flood risk management have emerged as useful methods (Kaczmarek et al., 1996). Non-structural measures include flood forecasting, warning systems, and floodplain zoning; green and blue infrastructures refer to waterways, green space, and trees useful for flood risk reduction (Perini, 2017). While these measures have become popular, engineering-oriented approaches have persisted and sometimes the actors supporting different methods are in conflict. For instance, the Water Framework Directive (WFD) adopted in 2000 by the European Union (EU) aims for integrated water management which promotes ecosystem conservation and wildlife habitat protection (European Commission 2019). Yet, one of the reasons why the implementation of the WFD has been slow until now is conflicts among stakeholders.

Struggles between different approaches to flood risk management can also be found in Germany and South Korea that have pursued contrasting policies for flood protection in the late 2000s. The decision-making processes of those policies were not without conflicts. In South Korea, environmental non-governmental organisations (NGOs) campaigned against the Four Rivers Project because the project included dam construction and dredging plans which have potentially harmful impacts on the ecosystem (KFEM, 2009; Kim, 2010). In the case of Germany, many natural floodplain and retention area extension projects which promote ecosystem conservation have faced opposition from local farmers who do not want their fields submerged during floods (Ehrenstein, 2013). I argue that those different approaches in conflicts stem from different discourses of water management introduced in the previous section.

In sum, flood risk management policy is a good case to examine discursive conflicts and struggles among stakeholders within the policy process. Although many angles exist when looking at these struggles, this study particularly focuses on the struggle between engineering-based and nature-based approaches to flood control in river basins. It is because these two contrasting approaches are the main difference found in the water management sector within the national climate change adaptation plans of Korea and Germany.

1.5 Research Design

1.5.1 Grounded Theory

This study is a piece of qualitative research that aims to uncover the components underpinning different policy processes in the field of water management, particularly related to adaptation to climate change. Given the relatively recent emergence of the climate change adaptation field, it seemed appropriate to take a grounded theory approach (Glaser & Strauss, 1967; Bryant & Charmaz, 2007; Mattoni, 2014).

Grounded theory seeks to establish a theory based on the constant comparison of empirical findings and theories throughout a research process (Mills et al., 2006; Flick, 2007). A distinct feature of grounded theory is that hypotheses are used as a means of showing the research direction (Creswell, 1998). Hypotheses are viewed as sensitising concepts that suggest where to focus on during the analysis (Mattoni, 2014). Mills et al. (2006) emphasise that the researcher has no preconceived ideas to test. Rather, essential issues emerge from interviews in the area of interest. By examining empirical data after collection, the researcher later can fill the sensitising concepts with meaning (Mattoni, 2014).

Another notable feature of grounded theory is that it follows an abductive research process (Mattoni, 2014; Blaikie, 2000). Abductive thinking differentiates itself from inductive or deductive reasoning. It does not start with a general theory as in deductive reasoning, nor does it generalise from a large number of observations (Gibbs, 2007). In abductive research, data gathering, data analysis, and concept building are not conducted in a linear manner. Instead, the three components are mutually connected and interact with each other simultaneously over the course of the research (Mattoni, 2014). Timmermans and Tavory (2012, p. 180) describe the process as "a dialectic". Fieldwork and data analysis are not separated from each other. Moreover, the theorising stage can influence the fieldwork (Mattoni, 2014).

Many data gathering methods used in grounded theory are common in social science practice in general. Although Glaser and Strauss (1967), who developed grounded theory, suggested that having an overall knowledge about the subject in question is sufficient to start the fieldwork, more recent researchers point out that recognising existing literature in the field is important (Mattoni, 2014). Strauss and Corbin (1990) added that diverse data, including the personal experience of researchers and previous studies, are valuable for grounded theory. Other data sources may include participant observation, in-depth interviews, focus groups, survey results, and historical documents. Diverse sources are essential to finding multiple viewpoints (Mattoni, 2014).

The data collected during this study was analysed following Mills et al. (2006) who divided the data analysis in grounded theory into at least three steps. The first is open coding, where the data (normally texts) is broken into small segments. The second stage is axial coding. This refers to grouping the segments into analytical categories. The last step of grounded theory data analysis is selective or focused coding. During this stage, the researcher creates more general categories at a higher level of abstraction and focuses on specific codes which appear to them as more important than others (Mattoni, 2014; Charmaz, 2006). After completing the three-step coding process, the results can be analysed in various ways. The data analysis methods applied in this study are explained in detail in Sect. 1.6.2.

1.5.2 Comparative Case Study

This study is based on a very similar system design to find the essential factors that explain differences in water management policy related to climate change adaptation. The most similar system design aims for explaining differences in the outcome for generally similar cases (Wagemann, 2014). For this study, South Korea and Germany were chosen because of their similarities in geographical and economic-political characteristics. The dependent variable is the water management policy within the first national adaptation strategies of the two countries, which were established in the late 2000s. More specifically, the unit of analysis is the Four Rivers Project in the case of South Korea and the National Flood Protection Programme (NFPP) in the case of Germany. Those policies (dependent variable) are notably different as South Korea indicated an engineering-based flood risk management project, while Germany emphasised re-naturalisation and nature-based solutions for the same policy goal. Such differences made the two cases stand out among industrialised countries. The differences between the cases will be described in detail in the cross-case analysis chapter (Chap. 6).

As this study looks into two cases in depth in a qualitative manner, it can be defined as a small-N comparative case study. This research design is based on the premise that effective analysis results from a careful examination of the cases, not from computations (Frendreis, 1983). A case study can be defined as "an intensive study of a single unit" (Gerring, 2004, p. 342). The aim of case studies is not merely to understand the case in focus. The implied assumption is that the causal mechanism found in one case, or a few cases in the same study, can possibly be applied to other similar cases (Hawkins, 2009). While case studies are poor at testing theory, they are usually better at constructing validity and generating new theories than large-N studies (Gerring, 2004).

If a case does not conform to a general tendency or has a significant impact on other cases, doing an in-depth case study can be beneficial (Hawkins, 2009). Although limits exist in the generalisation of conclusions, making hypothesis testing difficult in a single or small-N case study, the in-depth investigation makes theory development possible. The generated theory can then be tested in other cases in the future (George & Bennett, 2005; Ragin & Rubinson, 2009). Besides theory generation, case studies are also useful in the identification of causal mechanisms and asymmetric theoretical arguments (Hawkins, 2009).

In the context of this book, the case of South Korea is an interesting subject for investigation using case studies. Many industrialised countries have set up national adaptation plans and strategies (UNFCCC, 2019). If narrowed down to water management, South Korea possesses a policy which is not common among other industrialised countries. While many countries have shifted from a heavy engineering approach for flood management towards nature-based, non-structural approaches, the South Korean government has implemented a hard engineering project entitled the Four Rivers Project to adapt to future climate change. The project mainly consists of building dams and dredging riverbeds in the four major rivers in Korea. When considering that not many industrial countries included dam construction in their climate change adaptation strategies, the South Korean case is 'rare enough' to warrant investigation through the case study method. Therefore, the Korean case is compared to Germany, which showed a relatively clear turn from technical methods towards natural methods in terms of flood risk management. In this study, the policy of Germany represents the general approach to river management in industrial countries since the 2000s (Gralepois et al., 2016).

As Gerring (2004) suggests, a case is a unit bounded by a particular time and space. The cases of this study are the national climate change adaptation strategies in South Korea and Germany, established in the late

2000s. However, the water management policies referred to in the adaptation strategies are the results of a longer period of development. Therefore, the temporal boundary is from the 1950s until the 2010s—an approximately 70-year time period.

1.5.3 Case Selection Criteria

1.5.3.1 Geographical Similarities

There are several reasons to choose the climate change adaptation strategies of South Korea and Germany as case studies for this book. First, the two countries share some geographical similarities, particularly in terms of hydrology. As the Intergovernmental Panel on Climate Change (IPCC) suggests, geographical contexts are one of the key factors determining the effects of climate change (IPCC, 2014).

While South Korea has a higher average yearly precipitation level than Germany, both countries are facing more frequent and extreme precipitation due to climate change (Korea Meteorological Administration, 2017; BMU/UBA, 2018; Kunz et al., 2017). Flood control is a priority in South Korea and Germany because they both are densely populated. South Korea has 530 people per square kilometre of land, and Germany has 237 people per square kilometre which is well above the average of Europe (World Bank, 2019; BMU/UBA, 2018).

The two countries generally have temperate seasonal climates with four distinct seasons, but the seasonality of floods differs from each other: South Korea has a high risk of floods in the summer season, while most parts of Germany have a higher risk of floods in the winter season (Korea Meteorological Administration, 2014; Beurton & Thieken, 2009). Nevertheless, the similarity between the two countries is that large-scale flooding has occurred in major river basins in both countries. In South Korea, the four major rivers are the Han, Nakdong, Geum, and Yeongsan, while in Germany, the Danube, Rhine, Elbe, Weser, and Oder are referred to as the major rivers.

1.5.3.2 Hydrological Management Similarities

Korea and Germany have heavily modified their rivers, constructing grey infrastructure. Since the 1960s, the Korean government has built dams for controlling floods, securing water, and generating hydropower. Korea is well known for its high number of dams, with more than 1300 large dams as of 2010 (Kang, 2012; International Commission on Large Dams, 2019). The pace of the dam development was slowed down in the late 1990s, but the basic structures of water management policy did not change, as shown in the case of the Four Rivers Project.

In Germany, until 200 years ago, floodplains, high peat moors, and wetland forests were mostly intact along rivers (Blackbourn, 2006). From the mid-eighteenth century, Germans transformed their rivers rapidly. Floodplains were reclaimed for farmland and industrial facilities and rivers were canalised for flood control and navigation. Hundreds of dams were built for hydropower (Blackbourn, 2006; Federal Government, 2015).

However, river management has changed in Germany since the 1980s when environmentalism became part of the social discourse of the country. The changed policy for river management as well as flood risk management is reflected in the national adaptation plan: it prioritises the natural adaptive capacity of rivers and promotes re-naturalisation. This policy change in Germany makes Korea and Germany especially interesting cases for comparison.

1.5.3.3 Economic Similarities

The two countries experienced intensive economic growth after the Second World War. This fact is worth noting because the degree and content of economic development influence environmental policy to a great extent (Shafik, 1994; Arrow et al., 1995).

The fast growth of the West German economy in the 1950s after defeat in the Second World War was referred to as the Miracle on the Rhine, or *Wirtschaftswunder* in German (Strauss et al., 2019). The strong emphasis on economic growth at that time left little room for environmental protection (Schreurs & Papadakis, 2020). South Korea was still one of the poorest countries in the world after the Second World War and the situation was exacerbated by the Korean War (1950–1953). Beginning in the 1960s, however, South Korea's economy grew extremely fast (Scitovsky, 1985), and South Korea ascended into the ranks of the world's richest countries by the mid-1990s. West Germany's rapid economic growth was the role model for the South Korean government. The term "Miracle on the Han River" was coined after the "Miracle on the Rhine" by political leaders in order to encourage South Koreans to achieve economic growth similar to the situation in Germany. Industrialisation based on heavy industry was the motto and Germany was the model to follow (Han, 2016; "Let us prepare", 1961).

During the fast growth period, pressure to maintain economic growth was high in both countries, particularly because of their unique circumstances. Both were divided countries and this led to competition across their borders. Before the reunification of Germany, both West and East Germany were engaged in economic and ideological competition. South Korea was in a very similar situation in relation to North Korea until the 1980s. It was believed that higher and faster economic growth would represent the superiority of their political systems. This economic growth imperative strongly affected environmental quality and conditions (Ministry of Environment, 2010; Jones, 1993). For example, although the chemical industry was one of the main reasons for water pollution in both countries, criticising pollution problems was regarded as taboo as the pollution was accepted as an unavoidable side-effect of fast economic growth (Jung et al., 2014; Wurzel, 2002; Weale, 1993).

Currently, the two countries are among the richest world economies. Germany became a member of the Organisation for Economic Co-operation and Development (OECD) in 1961 and South Korea in 1996. The member countries of the OECD are industrialised countries which have a relatively high-income level. In 2017, Germany ranked 6th in terms of gross domestic product (GDP), while South Korea was 15th among 230 countries (The World Factbook, 2017).

1.5.3.4 Political System Similarities

Another factor that influences environmental policy is the political system (Arrow et al., 1995). Here the comparison is somewhat complicated as there have been political system changes in both countries.

From 1949, West Germany had a democratic government, whereas East Germany was governed by a single party (Socialist Unity Party of Germany). Since the reunification in 1990, however, Germany has had a stable democratic system. It is a federal state which consists of 16 states (*Länder*) with a parliamentary democracy. Legislative powers are assigned to the federal parliament (*Bundestag*) as well as state parliaments and the members of the parliaments are elected through regular elections (Facts about Germany, 2021). The federal government system has influenced German environmental politics as the legislation and implementation of relevant policies have been *Länder*'s responsibilities. More details of how federalism has affected water management will be presented in Sect. 5.5.1.

While national-level policies are crucial in water management, in Germany's case it is also necessary to look at developments in the EU. As an EU member state, Germany must conform to EU directives, including the Water Framework Directive and Floods Directive. Although the multi-level political system of the EU reduces the room for decision-making at the national and sub-national levels, nation-states are still crucial for environmental politics in Europe (Uekötter, 2014).

South Korea's democratisation process was rockier than Germany's until the 1980s. After its independence from Japanese colonisation in 1945 and the Korean War (1950–1953), two military authoritarian regimes of Park Chung Hee and Chun Doo Hwan ruled the country between 1961 and 1987. Persistent movements for democracy ended the Chun Doo Hwan regime leading to democratisation.

South Korea is a presidential democracy. The president is elected by popular vote every five years. The parliament consists of directly elected members (84%) and members elected through proportional representation (16%) (Korea.net, 2020). One of the distinctive features of the Korean political system is that power is highly concentrated at the national level. The president can exert power over environmental policy, including water management. This feature will be further explained in Sect. 4.5.2.

Although the political changes of the two countries had different pathways and they have different political system, the levels of democracy in the two countries today are similar. The Democracy Index of the Economist Intelligence Unit (EIU) places Germany 13th and South Korea 20th. According to the EIU, both have relatively high levels of democracy defined by many indicators such as high levels of political participation, free and fair electoral processes, and civil liberties (EIU, 2018).

Given the similarities in geographical, hydrological, economic, and political terms, a most similar system design can be used in exploring their different adaptation policies (Przeworski & Teune, 1970). This book explores the factors that explain Germany's turn to re-naturalisation politics while South Korea continues to stick to more technical solutions.

1.5.3.5 Research Gap in Comparative Environmental Politics of Germany and South Korea

There have been a few comparisons of the environmental policies of South Korea and Germany with most of the studies conducted by Korean researchers (e.g., Kim & Mayer, 2013; Yeom, 2014; Han & Lee, 2006).

In comparison with the rich literature on German environmental policies and politics, South Korean environmental policies have not gained much attention. The Korean case is not commonly addressed in the global water and adaptation literatures. There are few analyses of Korean environmental policies from a historical or political perspective (e.g., Lim & Tang, 2002; Choi et al., 2017) and almost none on Korean flood management policy in the international literature. This book's empirical case study of water management and climate change adaptation in Korea will therefore contribute to filling a research gap in the environmental politics literature.

Within Korean academia, adaptation to climate change is a new field; not many analyses have been conducted on the subject. The Four Rivers Project, the biggest adaptation project implemented in South Korea so far, has received some attention, but primarily from legal (Kim, 2011), civil engineering (Park, 2009), communications (Yun & Lee, 2010; Jeon & Kim, 2012), environmental policy process (Kim, 2010; Lee, 2010), and economic perspectives (Hong, 2008). Its integration into the country's climate change adaptation policies has not been seriously researched. This research will thus contribute to expanding the discussion on adaptation policies in South Korea with a policy process analysis perspective. The comparison will shed light on the power relationship of actors within climate change adaptation.

1.6 Research Procedure

1.6.1 Data Collection

Case studies typically combine different methods which could be qualitative, quantitative, or both (Eisenhardt, 1989). Triangulation refers to the use of more than one method for the same purpose. It can be performed during different stages and parts of the research. For example, data sources, investigators, theories, and methods can all be triangulated (Flick, 2007; Denzin, 1970). The advantage of triangulation is that a phenomenon can be shown from different viewpoints or different levels (Flick, 2007). Different views can be collected through different kinds of data, which helps the researcher to understand the broader picture (Glaser & Strauss, 1967). Thus, the quality of research can be enhanced through this technique.

This research combined three different methods in order to triangulate between them (Flick, 2007); in-depth interviews, participant observation, and a review of documents, including media, were used for data collection. The selection of different methods followed suggestions for triangulation given in methodology literature. For example, Flick (2007) suggested selecting one method which aims to understand subjective meanings and social background and another to analyse viewpoints and meanings beyond the present situation. He calls this a systematic triangulation of perspectives. Fielding and Fielding (1986) suggest a similar approach when combining more than one method. They recommend that at least one method should be used that helps to understand structural aspects of the problem and at least one for understanding the meaning of these aspects to the actors involved (Fielding & Fielding, 1986).

This study started with initial document analysis, mainly regarding adaptation policies with a focus on water management, in order to perform a structural analysis of the issue. Subsequently, the two cases were selected as the subjects for the comparative case study as they showed similarities in many aspects but significant differences in adaptation policies. Following the case selection, interviewing was chosen as the primary method for data collection as this method helps understanding different views of the actors. Besides this data source, the observation of related events (conferences, workshops, and a film festival) and media analysis provided useful data from diverse perspectives.

1.6.1.1 Semi-Structured Interview

Semi-structured interviews were conducted between February 2017 and January 2020 to collect the primary data. The interviewee selection was based on three criteria: participation, responsibility, and expertise in adaptation or water management policy processes. Interviewees were chosen from academia, government officials, and non-governmental actors in civil organisations or journalism (Huntjens et al., 2011).

The interviewees were first identified through relevant sources including media coverage, government documents, grey literature, and academic articles (Fischer, 2003). After initial interviews in the fieldwork based on purposeful sampling, snowball techniques were used to identify the next interviewees and broaden the interview sample (Mattoni, 2014). Interviewees recommended other relevant actors or experts who could provide interviews for the research. At the same time, the diversity of interviewees in different positions and roles was taken into account in order to build robust data.

The interviews were conducted both in South Korea and in Germany. Twenty-seven interviews were conducted in Korea and 20 interviews in Germany. In addition, seven interviews were conducted in the

Netherlands. Historically, the water management policy and technology of the Netherlands have influenced both countries. Germany and the Netherlands share several river basins, and policy transfer between the two countries often occurs at various levels including the EU. Also, Dutch technology and policy ideas of land reclamation and wetland restoration have historically influenced the Korean coastal management policy (Kang et al., 2022). Therefore, interviews with relevant experts in the Netherlands were helpful to understand the discursive and institutional dynamics of the policymaking process in water management. The list of interviewees is in Appendix 1. For confidentiality, the names of interviewees were anonymised when the interviewees requested.

The guiding questions were sent to most of the interviewees before the interviews. The interview questions were organised in order to elicit two different kinds of responses—narratives and statements (Kvale, 2007). Flick (2007) links narrative to narrative-episodic knowledge and statements to semantic conceptual knowledge. While narrative-episodic knowledge is related to situations, context, and personal opinion, semantic conceptual knowledge is about concepts, definitions, and relations (Flick, 2007).

The interviews in South Korea were conducted in Korean and the questions were translated from English to Korean. In Germany and the Netherlands, the interviews were conducted in English and the questions were not translated. All the interviews were recorded and transcribed.

1.6.1.2 Document Review (Primary Document Analysis)

Besides semi-structured interviews, relevant data from media, government documents, NGO statements, and other grey literature were collected through online search and library database search.

A media review was conducted for examining discourse change over time. As Boykoff (2011) states, "media messages are critical inputs to what becomes public discourse" (p. 3). A keyword search was undertaken in the digital archives of two mass media publications for each case. 'Floods' was the main keyword used and the time frame was from 1945 to 2019. In Germany, *Der Spiegel* and *Frankfurt Allgemeine Zeitung* were reviewed and in the Korean case, the *Kyunghyang Shinmun* and *Dong-a Ilbo* were chosen. The media were selected for two reasons: The selected publications have archives accessible online without charge and they are major newspapers well-known in respective country.

Other data sources comprise the websites and reports of national governments, sub-national governments, NGOs, international organisations, research institutes, and private companies. The primary search tool was a keyword search on the Internet. Some interviewees also provided printed documents during the interviews. The primary documents collected in various ways contained essential information for the analysis.

1.6.1.3 Observation

Another method used for collecting data was observation. Public events including conferences, workshops, talks, and a film festival were the main venues for observation. The information about relevant events were gathered from regular search on the Internet and also through interviewees. In South Korea, two workshops which dealt with water management policy were attended. Meanwhile, a film festival organised by an environmental NGO, and a talk event were attended in Germany. In addition, academic conferences related to environmental politics including water management were attended to collect secondary data. The list of attended events for this study can be found in Appendix 2. The data gained from these observation activities were not used as key data but helped to understand the background of the cases and to extend the interviewee pool.

1.6.2 Data Analysis

The collected data were analysed in two ways—coding and process tracing. The two analytical methods were combined to understand the policy process in South Korea and Germany better. Data handling and interpretation are necessary for data analysis (Gibbs, 2007). While coding is a useful tool for handling data and can provide information on actors' discourses, process tracing is useful in understanding the historical background of changes in dominant social discourses and institutions. Using the two analytical methods together, this study aims to generate a hypothesis on policy process in the climate change adaptation sector, particularly regarding water management.

1.6.2.1 Coding and Narrative Analysis

In this book, the data was coded with the support of MAXQDA, a package of computer-assisted qualitative data analysis software. Codes are the meaning that the researcher gives to the data (Gibbs, 2007). In text or other data forms, one or more passages can be identified in a code.

Coding is a widely used method both in quantitative and qualitative data analyses. However, the goals of coding in quantitative and qualitative analyses are different. While coding in quantitative analysis aims for counting the categorised responses to survey questions and assigning numbers to survey answers, coding in qualitative analysis is about organising, managing, and interpreting the data collected (Gibbs, 2007). As Saldaña (2016) states, coding aims to find meaningful information from data and the process, as well as the result, of coding reflect the researcher's worldview.

There are various ways to code data introduced by scholars. One way in grounded theory is to code in three steps: open coding, axial coding, and selective coding (Strauss & Corbin, 1990; Mills et al., 2006). While open coding means breaking the data (typically texts) into small segments, axial coding refers to grouping the segments by analytical category. Finally, in selective or focused coding, the researcher makes more general categories at a higher level of abstraction and focuses on a few selected categories which appear to them to be more critical than others (Mattoni, 2014; Charmaz, 2006). Gibbs (2007) also recommends three levels of coding: descriptive, categorising, and analytic. Meanwhile, Saldaña (2016) categorises diverse coding practices into two groups: first-cycle and second-cycle coding.

Although the suggested categories by different scholars have different names, there is a common feature among the categories: at the beginning, coding is close to the text and then gradually moves to more analytic and abstract concepts. As Saldaña (2016) states, coding at the later stage requires analytic skills such as classifying, prioritising, integrating, synthesising, abstracting, conceptualising, and theory building. Gibbs (2007) recommends moving away from descriptive codes to theoretical or analytical codes over the process.

In this study, coding started with open coding for the first cycle. The codes were more descriptive than analytical. Without a given list, open coding was conducted for all the interviews conducted in South Korea and Germany. In vivo codes were also created by adopting the concepts and words used by the interviewees (Glaser & Strauss, 1967). After the initial coding process, 930 codes were obtained. They were classified into 134 categories in the second cycle, axial coding.

After the first and second cycle coding, code mapping was undertaken for the third step. For this step, some codes were selected and mapped because of their significance in the analytical framework (discursive and institutional analyses). The Creative Coding (MAXMaps) function in the software MAXQDA was used for this step.

Lastly, narratives that appeared in the interviews were also analysed during the coding process. Storytelling is "one of the fundamental ways" that people express their understanding of the world (Gibbs, 2007, p. 76). The narrative analysis particularly focused on interviewees' stories which were closely linked to the codes selected in the third cycle coding process. Some of the interviewees' narratives are introduced in the case study chapters (Chaps. 4 and 5) as well as the cross-case chapter (Chap. 6).

1.6.2.2 Process Tracing

The coded data of this study were analysed using process-tracing methods. Process tracing is one of the fundamental methods for a within-case analysis (Hawkins, 2009). Originally from cognitive psychology, the term 'process tracing' refers to examining steps in an inference process in order to find out how the process occurs (Bennett & Checkel, 2015). Bennett and Checkel (2015) explain that the goal of the process-tracing method is to reveal the causal chain and causal mechanism between independent variables and the outcome of the dependent variable. To that end, diverse sources such as histories, archival documents, and interview transcripts are utilised (George & Bennett, 2005).

The main observation point during a process-tracing analysis is to find "the chain of events" which constitutes the hypothesised causal mechanism (Hawkins, 2009). Unlike the usual assumption in a hypothesis test, the events between the independent variable(s) and the dependent variable are not regarded as mere 'intervening variables' which transmit the causal force. Instead, process tracing focuses on the evidence of sequences and conjunctures of events, which either have independent effects on the dependent variable or interact with the independent variables (Bennett & Checkel, 2015). Therefore, it is also called "causal process observation" (Collier, 2011, p. 823). As with other methods, identifying causal mechanisms is a challenging task even with process tracing (Hawkins, 2009). Bennett and Checkel (2015) highlight diligent search for alternative explanations and openness to the potential biases of sources. Notably, before starting an analysis with process-tracing methods, it is useful to find out as many facts and sequences as possible within the case which should be true if alternative explanations are true (Bennett & Checkel, 2015).

Process tracing can be used for both testing an established theory and developing a new one. When process tracing is used for inductive research, hypotheses are generated based on the evidence found in a case (Bennett & Checkel, 2015). In this study, process tracing is used for establishing a hypothesis using the evidence provided by interviews and other sources. The use of process tracing enabled the identification of critical junctures, and discourse and institutional changes. The main focus of the analysis step that applied the process-tracing method was on the causal mechanism between critical events to water management policies (potential critical junctures) and policy changes. In this way, this study aimed to clarify how and when critical junctures influence water management and climate change adaptation policy processes.

1.7 Overview of the Book's Structure

This book consists of seven chapters. Chapter 1 briefly explains the background of this study before presenting the research question. It also addresses the relevance of the research question in the context of climate change adaptation and water management literature. In addition, Chap. 1 introduces the research design and methods including the criteria used for the case study selection.

Chapter 2 presents the analytical framework adopted for this study. This chapter presents a literature review of policy process analysis, discourse analysis, and institutional analysis in the context of environmental politics, as these are the bodies of theoretical literature that support this thesis. First, the main approaches for policy process analysis are reviewed and the selected framework, discursive institutionalism, is explained. Although at the early stage the research began by treating discursive analysis and institutional analysis separately, the interdependent relationship between discourse and institution emerged during the data analysis process. Vivien Schmidt's discursive institutionalism was therefore adopted as the main analytical framework. Schmidt (2010) provides a broad definition of discursive institutionalism, mentioning that it refers to political science works that deal with the role of discursive interactions in institutional changes. Based on this definition, this book investigates the two cases in three steps: (1) analysis using the concepts of path dependence and critical junctures,[5] (2) identification of social, programmatic, and policy discourses, and (3) examination of the dynamic interactions between discourse and institutions related to water management and climate change adaptation in the two countries.

The background to this study, including the challenges of climate change, adaptation and water systems in general is explored in Chap. 3.

[5] Path dependence and critical junctures are two key analytical concepts of historical institutionalism, which is one of four new institutionalist approaches (rational choice, sociological, historical, and discursive institutionalism. This study applies discursive institutionalism for its analytical framework, but it borrows some analytical elements from historical institutionalism because discursive institutionalism is an extensive concept that includes all approaches that take account of ideas and the interactions within new institutionalism. See Sect. 2.3.3 for more explanations of discursive institutionalism in historical institutionalism tradition.

The scientific evidence for a changing climate is presented, and the impacts of climate change on social-ecological systems is outlined as justification for actions on adaptation. Following this, the chapter presents further explanation as to the meaning of adaptation and the development history of adaptation policy institutions. A literature review of different discourses on adaptation is then conducted and explained, and Pelling's (2011) three types of adaptation (resistance, transition, and transformation) are introduced. These types of adaptation are categorised by the scope and range of adaptation and they are used together with discursive and institutional analyses in the cross-case analysis. Following this background on adaptation, the impacts of climate change on water, and policy reactions to such changes are presented. Finally, four discourses regarding water management are reviewed followed by a historical overview of institutional changes in the sector. The discourses and the historical changes of water management provide useful background information for comparing the empirical findings of the two case studies in Chap. 6.

Chapter 4 deals with the first case, South Korea. It starts with past records and future projections of climate change, and then deals with the impact of climate change on water management, forming the background information for Korea's adaptation policy. This is followed by a description of Korean policies relevant to adaptation and water management. First of all, the Four Rivers Project is explained in detail followed by the description of the National Adaptation Plan and the Green Growth scheme, which influenced the project. As well as adaptation policies, a brief history of water policy as implemented from 1960 until 2019 is presented, divided into the areas of water quality management and water quantity management. After the description of the adaptation and water policies, which are the most influential policies to flood risk management (the Four Rivers Project), relevant social discourses in South Korea such as developmentalism, techno-centrism, and environmentalism are analysed. Finally, the institutional setting of Korean environmental policy is analysed. The divided water management institution is one of the major characteristics of Korean water policy. Additionally, developmental state institutions which have promoted state-guided, high economic growth (Johnson, 1982), and strong presidency have contributed to the strong path dependency of flood risk management. A number of important events and accidents have affected the development of water management policy, but, overall, the progression of water policy has not substantially changed.

Chapter 5 follows the same structure as the previous chapter as it handles the second case, Germany. Like in South Korea, scientific evidence shows that Germany's climate has changed historically and is continuing to do so. The ways that water management is affected by the changing climate are described in this section. In regard to adaptation policy, supranational and national strategies are explained with a particular focus on water management. The structure and contents of water policy in Germany are then outlined. Because German politics is influenced by multi-level governance, relevant policies such as the EU Adaptation Strategy, the Water Framework Directive, and the Floods Directive are introduced. In addition, because river management is the main focus of this dissertation, adaptation plans in the major river basins are described in this section. After the description of the key policies related to flood risk management, this chapter investigates the institutions and social discourses relevant to the adaptation and water policies. In terms of discourse, optimism for technology and environmentalism have had the biggest influence over water management policy. This chapter ends by discussing institutions related to water management such as federalism and participatory governance. Through the media report analysis, critical junctures which changed German water policy are shown in the last section. The discourse and institutional analyses for each case presented in Chaps. 4 and 5 are discussed from a comparative perspective in Chap. 6.

In Chap. 6, the two cases of Germany and South Korea are compared through discursive institutionalism presented in Chap. 2. First, the ways in which social discourses have shaped the discourses on water management (programmatic discourse) and flood risk management (policy discourse) are compared. The significant difference between the two cases is that environmentalism is stronger in the political decision-making processes in Germany. As a result, a conservation-ecosystem service discourse became dominant in the water management sector, including flood control. The differences between the formal and informal institutions of South Korea and Germany are also a critical factor which resulted in different flood risk management. A strong presidency and centralised decision-making process characterise Korean environmental politics, whereas participatory and multi-level governance is viewed as a critical element of German environmental policy. After the comparison of the cases, the analysis moves on to discuss how discourse and institutions interact with each other in the policy process, based on the empirical evidences. Discourse change can be either facilitated or constrained by institutional settings, while institutions can also change through discursive changes. This finding is consistent with earlier findings within the school of discursive institutionalism.

The conclusion (Chap. 7) presents a summary of the argument with a reflection on this research project as a whole. It also discusses the implications of this study in the climate change adaptation field as well as environmental politics related to South Korea and Germany.

References

Alexander, M., Priest, S., & Mees, H. (2016). A framework for evaluating flood risk governance. *Environmental Science & Policy, 64*, 38–47. https://doi.org/10.1016/j.envsci.2016.06.004

Arrow, K., Bolin, B., Costanza, R., Dasgupta, P., Folke, C., Holling, C. S., et al. (1995). Economic growth, carrying capacity, and the environment. *Ecological Economics, 15*(2), 91–95. https://doi.org/10.1016/0921-8009(95)00059-3

Barriopedro, D., Fischer, E. M., Luterbacher, J., Trigo, R. M., & García-Herrera, R. (2011). The hot summer of 2010: Redrawing the temperature record map of Europe. *Science, 332*(6026), 220–224. https://doi.org/10.1126/science.1201224

Bauer, A., & Steurer, R. (2015). National Adaptation Strategies, what else? Comparing adaptation mainstreaming in German and Dutch water management. *Regional Environmental Change, 15*(2), 341–352. https://doi.org/10.1007/s10113-014-0655-3

Bennett, A., & Checkel, J. T. (Eds.). (2015). *Process tracing*. Cambridge University Press.

Berrang-Ford, L., Ford, J. D., & Paterson, J. (2011). Are we adapting to climate change? *Global Environmental Change, 21*(1), 25–33. https://doi.org/10.1016/j.gloenvcha.2010.09.012

Beurton, S., & Thieken, A. H. (2009). Seasonality of floods in Germany. *Hydrological Sciences Journal, 54*(1), 62–76. https://doi.org/10.1623/hysj.54.1.62

Blackbourn, D. (2006). *The conquest of nature. Water, landscape and the making of modern Germany*. Norton.

Blaikie, N. (2000). *Designing social research*. Polity Press.

BMU. (2017). *Fragen und Antworten zum Nationalen Hochwasserschutzprogramm*. Edited by BMU. Retrieved December 28, 2019, from https://www.bmu.de/themen/wasser-abfall-boden/binnengewaesser/hochwasservorsorge/nationaler-hochwasserschutz/hochwasserschutzprogramm

BMU/UBA. (2018). *Water resource Management in Germany. Fundamentals, pressures, measures*. UBA.

Bosomworth, K. (2018). A discursive–institutional perspective on transformative governance: A case from a fire management policy sector. *Env Pol Gov, 28*(6), 415–425. https://doi.org/10.1002/eet.1806

Boykoff, M. T. (2011). *Who speaks for the climate? Making sense of media reporting on climate change*. Cambridge University Press.

Bryant, A., & Charmaz, K. (Eds.). (2007). *The SAGE handbook of grounded theory*. SAGE Publications.

Bullock, A., Cosgrove, W., van der Hoek, W., & Winpenny, J. (2009). Getting out of the box–linking water to decisions for sustainable development. In World Water Assessment Programme (Ed.), *The United Nations world water development report 3: Water in a changing world* (pp. 3–24). UNESCO & Earthscan.

Charmaz, K. (2006). *Constructing grounded theory*. SAGE Publications.

Choi, I., Shin, H., Nguyen, T., & Tenhunen, J. (2017). Water policy reforms in South Korea: A historical review and ongoing challenges for sustainable water governance and management. *In Water, 9*(9), 717. https://doi.org/10.3390/w9090717

Collier, D. (2011). *Understanding Process Tracing. APSC, 44*(04), 823–830. https://doi.org/10.1017/S1049096511001429

Creswell, J. W. (1998). *Qualitative inquiry and research design: Choosing among five traditions*. Sage Publications.

Denzin, N. K. (1970). *The research act*. A theoretical introduction to sociological methods.

Depietri, Y., & McPhearson, T. (2017). Integrating the Grey, Green, and blue in cities: Nature-based solutions for climate change adaptation and risk reduction. In N. Kabisch, H. Korn, J. Stadler, & A. Bonn (Eds.), *Nature-based solutions to climate change adaptation in urban areas. Theory and practice of urban sustainability transitions*. Springer.

Deutschländer, T., & Mächel, H. (2017). Temperatur inklusive Hitzewellen. In G. P. Brasseur, D. Jacob, & S. Schuck-Zöller (Eds.), *Klimawandel in Deutschland* (pp. 48–56). Springer.

Dieperink, C., Hegger, D. L. T., Bakker, M. H. N., Kundzewicz, Z. W., Green, C., & Driessen, P. P. J. (2016). Recurrent governance challenges in the implementation and alignment of flood risk management strategies: A review. *Water Resources Management, 30*(13), 4467–4481. https://doi.org/10.1007/s11269-016-1491-7

Driessen, P. P. J., Hegger, D. L. T., Bakker, M. H. N., van Rijswick, H. F. M. W., & Kundzewicz, Z. W. (2016). Toward more resilient flood risk governance. *E&S, 21*(4). https://doi.org/10.5751/ES-08921-210453

Dupuis, J., & Biesbroek, R. (2013). Comparing apples and oranges: The dependent variable problem in comparing and evaluating climate change adaptation policies. *Global Environmental Change, 23*(6), 1476–1487. https://doi.org/10.1016/j.gloenvcha.2013.07.022

Eakin, H., & Lemos, M. C. (2006). Adaptation and the state: Latin America and the challenge of capacity-building under globalization. *Global Environmental Change, 16*, 7–18.

Ehrenstein, C. (2013, July 25). Deiche oder Auen? Das ist hier die Frage. *Welt*. https://www.welt.de/politik/deutschland/article118390294/Deiche-oder-Auen-Das-ist-hier-die-Frage.html

Eisenhardt, K. M. (1989). Building theories from case study research. *The Academy of Management Review, 14*(4), 532–550.

EIU. (2018). *Democracy index 2017: Free speech under attack*. The Economist Intelligence Unit Limited.

Eriksen, S., Nightingale, A. J., & Eakin, H. (2015). Reframing adaptation. The political nature of climate change adaptation. *Global Environmental Change, 35*, 523–533. https://doi.org/10.1016/j.gloenvcha.2015.09.014

European Commission. (2009). *White paper: Adapting to climate change: Towards a European framework for action*.

European Environment Agency. (2014). *National adaptation policy processes in European countries–2014*. Edited by European Environment Agency.

Facts about Germany. (2021). *Political system*. Retrieved December 23, 2021, from https://www.tatsachen-ueber-deutschland.de/en/politics-germany/political-system

Federal Government. (2015). *Fortschrittsbericht zur Deutschen Anpassungsstrategie an den Klimawandel.* The Federal Government.
Fielding, N., & Fielding, J. (1986). *Linking data.* SAGE.
Fischer, F. (2003). *Reframing public policy.* Oxford University Press.
Firus, K., Fleischhauer, M., Greiving, S., Grifoni, P., & Stickler, T. (2011). *Planning and implementing communication and public participation processes in flood risk management.* Retrieved March 27, 2020, from https://issuu.com/patri-cnr/docs/planning_comm_particip_floodrisk
Flick, U. (2007). *Managing quality in quantitative research.* SAGE.
Ford, J. D., Berrang-Ford, L., & Paterson, J. (2011). A systematic review of observed climate change adaptation in developed nations. *Climatic Change, 106*(2), 327–336. https://doi.org/10.1007/s10584-011-0045-5
Frendreis, J. P. (1983). Explanation of variation and detection of covariation. *Comparative Political Studies, 16*(2), 255–272. https://doi.org/10.1177/0010414083016002006
Fünfgeld, H. & McEvoy, D. (2014). Frame Divergence in climate change adaptation policy. Insights from Australian local government planning. Environ Plann C Gov Policy, 32 (4), 603–622. DOI: https://doi.org/10.1068/c1234.
Füssel, H. M. (2007). Adaptation planning for climate change: Concepts, assessment approaches, and key lessons. *Sustainability Science, 2*, 265–275.
George, A. L., & Bennett, A. (2005). *Case studies and theory development in the social sciences.* MIT Press.
Gerring, J. (2004). What is a case study and what is it good for? *The American Political Science Review, 98*(2), 341–354.
Gibbs, G. (2007). *Analyzing qualitative data.* SAGE.
Glaser, B. G., & Strauss, A. L. (1967). *The discovery of grounded theory.* Aldine Publishing.
Global Water Partnership. (2011). *What is IWRM?* Retrieved February 8, 2020, from https://www.gwp.org/en/GWP-CEE/about/why/what-is-iwrm/
Gralepois, M., Larrue, C., Wiering, M., Crabbé, A., Tapsell, S., Mees, H., Ek, K., & Szwed, M. (2016). Is flood defense changing in nature? Shifts in the flood defense strategy in six European countries. *Ecology and Society, 21*(4), 37.
Greiving, S., & Fleischhauer, M. (2012). National Climate Change Adaptation Strategies of European states from a spatial planning and development perspective. *European Planning Studies, 20*(1), 27–48. https://doi.org/10.1080/09654313.2011.638493

Grothmann, T., & Patt, A. (2005). Adaptive capacity and human cognition: The process of individual adaptation to climate change. *Global Environmental Change, 15*(3), 199–213. https://doi.org/10.1016/j.gloenvcha.2005.01.002
Hallegatte, S. (2009). Strategies to adapt to an uncertain climate change. *Global Environmental Change, 19*, 240–247.
Han, Y. (2016). *Miracle from the Rhein to the Han River: Heavy industrialisation of South Korea and its social consequences under Park Chung Hee*. PhD. Lancaster University. Retrieved October 16, 2019, from https://eprints.lancs.ac.uk/id/eprint/79039/
Han, Y.-H., & Lee, T.-G. (2006). Comparative study of the system for decentralized rainwater Management in Korea and Germany. *Journal of the Korean Institute of Landscape Architecture, 34*(4), 84–95. Retrieved March 22, 2020, from http://ocean.kisti.re.kr/downfile/volume/kila/JGHHBR/2006/v34n4s117/JGHHBR_2006_v34n4s117_84.pdf
Hawkins, D. (2009). Case Studies. In T. Landman & N. Robinson (Eds.), *Sage handbook of comparative politics* (pp. 50–66). SAGE Publications.
Hegger, D. L. T., Driessen, P. P. J., Wiering, M., van Rijswick, H. F. M. W., Kundzewicz, Z. W., Matczak, P., et al. (2016). Toward more flood resilience: Is a diversification of flood risk management strategies the way forward? *E&S, 21*(4). https://doi.org/10.5751/ES-08854-210452
Hong, J. (2008). *Is the Korean Grand Canal project economically feasible?* Korean Society of Environment and Ecology.
Huitema, D., & Meijerink, S. (2009). *Water policy entrepreneurs*. Edward Elgar.
Huitema, D., & Meijerink, S. (2010). Realizing water transitions. The role of policy entrepreneurs in water policy change. *Ecology and Society, 15*(2).
Huntjens, P., Pahl-Wostl, C., Rihoux, B., Schlüter, M., Flachner, Z., Susana, N., et al. (2011). Adaptive water management and policy learning in a changing climate. A formal comparative analysis of eight water management regimes in Europe, Africa and Asia. *Env. Pol. Gov., 21*(3), 145–163. https://doi.org/10.1002/eet.571
Ingram, H. (2011). Beyond universal remedies for good water governance. In A. Garrido & H. M. Ingram (Eds.), *Water for food in a changing world* (pp. 241–261). Routledge.
International Commission on Large Dams. (2019). *Number of dams by country members*. Retrieved January 3, 2020, from https://www.icold-cigb.org/article/GB/world_register/general_synthesis/number-of-dams-by-country-members
IPCC. (2014). *Climate change 2014: Impacts, adaptation and vulnerability, volume 2, regional aspects*. Cambridge University Press.

IPCC. (2021). In V. Masson Delmotte, P. Zhai, A. Pirani, S. L. Connors, C. Péan, S. Berger, N. Caud, Y. Chen, L. Goldfarb, M. I. Gomis, M. Huang, K. Leitzell, E. Lonnoy, J. B. R. Matthews, T. K. Maycock, T. Waterfield, O. Yelekçi, R. Yu, & B. Zhou (Eds.), *Summary for policymakers. In climate change 2021: The physical science basis. Contribution of working group I to the sixth assessment report of the intergovernmental panel on climate change.* Cambridge University Press. In Press.

Jeon, H. J., & Kim, J. S. (2012). Effects of message sidedness on credibility of the message and attitude toward the conflict issue. *Journal of Public Relations, 16*(2), 5–31.

Johnson, C. (1982). *MITI and the Japanese miracle.* Stanford University Press.

Jones, M. E. (1993). Origins of the east German environmental movement. In German Studies Review, 16 (2), p. 235. DOI: https://doi.org/10.2307/1431647.

Juhola, S., Keskitalo, E., Carina, H., & Westerhoff, L. (2011). Understanding the framings of climate change adaptation across multiple scales of governance in Europe. *Environmental Politics, 20*(4), 445–463. https://doi.org/10.1080/09644016.2011.589571

Jung, H., Lee, K., Jung, H., Kim, T., Chu, J., & Jeon, D. (Eds.). (2014). *Korean environmental policy.* Environment and Civilization.

Kaczmarek, Z., Amell, N. W., & Stakhiv, E. Z. (1996). Water resources management. In R. T. Watson, M. C. Zinyowera, & R. H. Moss (Eds.), *Climate change 1995. Impacts, adaptations and mitigation of climate change: Scientific-technical analyses. Contribution of working group II to the second assessment report of the intergovernmental panel on climate change.* 469, Cambridge University Press.

Kaika, M. (2003). The water framework directive: A new directive for a changing social, political and economic European framework. *European Planning Studies, 11*(3), 299–316. https://doi.org/10.1080/09654310303640

Kang, Y. H., Dieperink, C., & Hegger, D. (2022). Policy translation and dynamics: The role of Dutch ideas in developing South Korea's coastal management policies. *Ocean & Coastal Management, 221.*

KFEM. (2009). *Ten truths about the four rivers project.* Retrieved April 23, 2020, from http://kfem.or.kr/?p=2357

Kim, C. S. (2010). An evaluation on 4 Major River restoration project: The probabilities of excluding policy errors. *Korean Public Administration Quarterly, 22*(4), 1041–1066.

Kang, J. (2012). *The history of dam building in Korea: Korea national committee on large dams.* Retrieved February 4, 2020, from http://www.kncold.or.kr/ds_imgs/sub04/40/266.pdf

Kim, H.-R. (2011). Review of water management Laws and Recommendations. *Public Land Law Review, 52.*

Kim, Y. S., & Mayer, M. (2013). A comparative study of the energy policy making of South Korea and Germany: With focused to hermeneutic methodology. *Journal of Korean Association for Comparative Government, 17*(1), 95–124.

Korea Meteorological Administration. (2014). *Korean climate change assessment report 2014.* KMA.

Korea Meteorological Administration. (2017). *Climate change projection report in the Korean peninsula for the new climate regime.* KMA.

Kunz, M., Mohr, S., & Werner, P. C. (2017). Niederschlag. In G. P. Brasseur, D. Jacob, & S. Schuck-Zöller (Eds.), *Klimawandel in Deutschland* (Vol. 42, pp. 57–66). Springer.

Korea.net. (2020). *Constitution and government.* Retrieved 2 January, 2022, from https://www.korea.net/Government/Constitution-and-Government/Executive-Legislature-Judiciary

Kvale, S. (2007). Doing interviews. *SAGE.*

Let us prepare the ground for prosperity by austerity and diligence. (1961, January 1). *Kyunghyang Shinmun.*

Lee, S.-H. (2010). Evaluation over the legitimacy of 4 Major River project of Korea with the viewpoint of principle of integrated water management. *ECO, 14*(1), 63–96.

Lesnikowski, A., Ford, J., Biesbroek, R., Berrang-Ford, L., & Heymann, S. J. (2016). National-level progress on adaptation. *Nature Clim Change, 6*(3), 261–264. https://doi.org/10.1038/NCLIMATE2863

Lim, J. H., & Tang, S.-Y. (2002). Democratization and environmental policy-making in Korea. *Governance, 15*(4), 561–582. https://doi.org/10.1111/1468-0491.00201

Lindseth, G. (2005). Local level adaptation to climate change: Discursive strategies in the Norwegian context. *Journal of Environmental Policy & Planning, 7*(1), 61–84. https://doi.org/10.1080/15239080500251908

Mattoni, A. (2014). The potentials of grounded theory in the study of social movements. In D. Della Porta (Ed.), *Methodological practices in social movement research* (pp. 21–42). Oxford University Press.

Mertz, O., Halsnaes, K., Olesen, J. E., & Rasmussen, K. (2009). Adaptation to climate change in developing countries. *Environmental Management, 43*(5), 743–752. https://doi.org/10.1007/s00267-008-9259-3

Meijerink, S., & Huitema, D. (2010). Policy entrepreneurs and change strategies: Lessons from sixteen case studies of water transitions around the globe. *Ecology and Society, 15*(2).

Mills, J., Bonner, A., & Francis, K. (2006). The development of constructivist grounded theory. *International Journal of Qualitative Methods, 5*(1), 25–35.

Milly, P. C. D., Betancourt, J., Falkenmark, M., Hirsch, R. M., Kundzewicz, Z. W., Lettenmaier, D. P., & Stouffer, R. J. (2008). Climate change. Stationarity is dead: Whither water management? *Science, 319*(5863), 573–574. https://doi.org/10.1126/science.1151915

Ministry of Environment. (2010). *Three decades of environmental policies in Korea*. Ministry of Environment.

Ministry of Land, Transport and Maritime Affairs. (2009). *Masterplan for four Rivers*. Ministry of Land, Transport and Maritime Affairs.

Mosello, B. (2015). *How to Deal with climate change?* Springer International Publishing.

Moser, S. C., & Ekstrom, J. A. (2010). A framework to diagnose barriers to climate change adaptation. *PNAS, 107*(51), 22026–22031.

Moss, T., & Newig, J. (2010). Multilevel water governance and problems of scale: Setting the stage for a broader debate. *Environmental Management, 46*, 1–6.

Nightingale, A. (2017). Power and politics in climate change adaptation efforts: Struggles over authority and recognition in the context of political instability. *Geoforum, 84*, 11–20. https://doi.org/10.1016/j.geoforum.2017.05.011

OECD. (2013). *Water and climate change adaptation*. IWA Publishing.

O'Gorman, P. A., & Schneider, T. (2009). The physical basis for increases in precipitation extremes in simulations of 21st-century climate change. *In Proceedings of the National Academy of Sciences of the United States of America, 106*(35), 14773–14777. https://doi.org/10.1073/pnas.0907610106

Pahl-Wostl, C. (2015). *Water governance in the face of global change*. Springer International Publishing.

Pahl-Wostl, C., Holtz, G., Kastens, B., & Knieper, C. (2010). Analyzing complex water governance regimes: The management and transition framework. *Environmental Science and Policy, 13*, 571–581.

Park, C.-K. (2009). Some problems of the Green growth policy and the project saving the four Major Rivers. *E&S, 83*, 117–360.

Pelling, M. (2011). *Adaptation to climate change*. Routledge.

Perini, K. (2017). Green and blue infrastructure in cities. In K. Perini & P. Sabbion (Eds.), *Urban sustainability and river restoration* (Vol. 14, pp. 1–9). John Wiley & Sons Ltd.

Pittock, J. (2011). *National climate change policies and sustainable water management conflicts and synergies. E&S* 16.

Plate, E. (2002). Flood risk and flood management. *Journal of Hydrology, 267*(1–2), 2–11.

Przeworski, A., & Teune, H. (1970). *The logic of comparative social inquiry*. John Wiley & Sons Ltd.

Ragin, C. C., & Rubinson, C. (2009). The distinctiveness of comparative research. In T. Landman & N. Robinson (Eds.), *Sage handbook of comparative politics* (pp. 13–34). SAGE Publications.

Risbey, J. S. (2008). The new climate discourse: Alarmist or alarming? *Global Environmental Change, 18*(1), 26–37. https://doi.org/10.1016/j.gloenvcha.2007.06.003

Saldaña, J. (2016). *The coding manual for qualitative researchers*. SAGE.

Schmidt, V. A. (2010). Taking ideas and discourse seriously: Explaining change through discursive institutionalism as the fourth 'new institutionalism'. *Eur. Pol. Sci. Rev., 2*(01), 1. https://doi.org/10.1017/S175577390999021X

Schreurs, M., & Papadakis, E. (2020). *Historical dictionary of the Green movement*. Rowman & Littlefield Publishers.

Scitovsky, T. (1985). Economic development in Taiwan and South Korea: 1965-81. *Food Research Institute Studies, 19*(3).

Scoones, I., Leach, M., & Newell, P. (2015). *The politics of green transformations*. Routledge.

Shafik, N. (1994). Economic development and environmental quality: An econometric analysis. *Oxford Economic Papers, 46*(4), 757.

Smith, L. A., & Stern, N. (2011). Uncertainty in science and its role in climate policy. *Philosophical Transactions. Series A, Mathematical, Physical, and Engineering Sciences, 369*(1956), 4818–4841. https://doi.org/10.1098/rsta.2011.0149

Smit, B., & Wandel, J. (2006). Adaptation, adaptive capacity and vulnerability. *Global Environmental Change, 16*, 282–292.

Strauss, A., & Corbin, J. M. (1990). *Basics of qualitative research: Grounded theory procedures and techniques*. SAGE Publications.

Strauss, G., Leyser, K. J. et al. (2019). *Germany*. Retrieved October 16, 2019, from https://www.britannica.com/place/Germany

Tanaka, S. K., Zhu, T., Lund, J. R., et al. (2006). Climate warming and water management adaptation for California. *Climatic Change, 76*, 361–387.

The World Factbook 2017. (2017). *Central intelligence agency*. Retrieved September 8, 2019, from https://www.cia.gov/library/publications/the-world-factbook/rankorder/2001rank.html

Thieken, A. H., Kienzler, S., Kreibich, H., Kuhlicke, C., Kunz, M., Mühr, B., et al. (2016). Review of the flood risk management system in Germany after the major flood in 2013. *E&S, 21*(2). https://doi.org/10.5751/ES-08547-210251

Timmermans, S., & Tavory, I. (2012). Theory construction in qualitative research. *Sociological Theory, 30*(3), 167–186. https://doi.org/10.1177/0735275112457914

Trenberth, K. E. (2011). Changes in precipitation with climate change. *Climate Research, 47*(1), 123–138. https://doi.org/10.3354/cr00953

Trottier, J. (2004). Introduction. In J. Trottier & P. Slack (Eds.), *Managing water resources past and present* (pp. 1–12). Oxford University Press.

Uekötter, F. (2014). *The greenest nation? A new history of German environmentalism*. The MIT Press.

UNFCCC. (2018). *Adaptation plans and strategies*. Retrieved January 21, 2020, from https://www4.unfccc.int/sites/NAPC/Pages/adaptation-plans-and-strategies.aspx

UNFCCC. (2019). *Seventh National Communications–Annex I*. Retrieved January 21, 2020, from https://unfccc.int/process-and-meetings/transparency-and-reporting/reporting-and-review-under-the-convention/national-communications-and-biennial-reports-annex-i-parties/seventh-national-communications-annex-i

Vogel, M. M., Zscheischler, J., Wartenburger, R., Dee, D., & Seneviratne, S. I. (2019). Concurrent 2018 hot extremes across northern hemisphere due to human-induced climate change. *Earth's Future, 7*(7), 692–703. https://doi.org/10.1029/2019EF001189

Wagemann, C. (2014). Qualitative comparative analysis (QCA). In D. Delta Porta (Ed.), *Methodological practices in social movement research* (pp. 43–66). Oxford University Press. https://doi.org/10.1093/acprof:oso/9780198719571.003.0003

Weale, A. (1993). Ecological modernisation and the integration of European environmental policy. In J. D. Liefferink, P. Lowe, & A. P. J. Mol (Eds.), *European integration and environmental policy*. Belhaven.

Wilby, R. L., & Dessai, S. (2010). Robust adaptation to climate change. *Weather, 65*(7), 180–185. https://doi.org/10.1002/wea.543

Wilhite, D. A., Sivakumar, M. V. K., & Pulwarty, R. (2014). Managing drought risk in a changing climate: The role of national drought policy. *Weather and Climate Extremes, 3*, 4–13. https://doi.org/10.1016/j.wace.2014.01.002

Wise, R. M., Fazey, I., Stafford Smith, M., Park, S. E., Eakin, H. C., van Archer Garderen, E. R. M., & Campbell, B. (2014). Reconceptualising adaptation to climate change as part of pathways of change and response. *Global Environmental Change, 28*, 325–336. https://doi.org/10.1016/j.gloenvcha.2013.12.002

World Bank. (2019). *Population density* (people per sq. km of land area). Retrieved January 3, 2020, from https://data.worldbank.org/indicator/EN.POP.DNST?year_high_desc=true

Wurzel, R. (2002). *Environmental policy-making in Britain, Germany and the European Union. The Europeanisation of air and water pollution control.* Manchester University Press.

Yeom, Kwanghee (2014). Comparative Study on the conflict in RE facility siting in Korea and Germany. In *Environmental law and policy* 13. Retrieved March 22, 2020, from http://cls.kangwon.ac.kr/data/file/sub41/1917250875_a00d539d_13_7_BFB0B1A4C8F1.pdf

Yun, S.-J., & Lee, D.-H. (2010). Agenda setting and frame of TV news about 4 Major Rivers project in Korea. *ECO, 14*(1), 7–62.

2

A Discursive Institutionalist's Approach to Policy Process in the Tradition of Historical Institutionalism

2.1 Policy Process Analysis

2.1.1 Different Approaches for Analysing Policy Process

The two major subjects within policy process analysis are policy change and stability (John, 2003; Cairney, 2013). As policy process is the core subject of this book, it is necessary to start with an explanation of what policy means in this study. There are several existing definitions of policy. When the concept is narrowed down to that of public policy, which is the focus of this book, Dye's (1984) definition is simple and clear: "whatever governments choose to do or not to do" (p. 1). The governments' intention of doing or not doing something can be supported through incentives and punishments (Lowi & Ginsberg, 1996). Public policies can take many forms, including statutes, laws, edicts, regulations, orders, and government projects and programmes (Weible, 2014; Fischer, 2003). In addition, commonly understood rules-in-use that influence people's

Y. h. Kang, *Climate Change Adaptation in River Management*, Palgrave Studies in Water Governance: Policy and Practice,
https://doi.org/10.1007/978-3-031-10486-2_2

behaviours in public affairs can also be regarded as public policy (Schneider & Ingram, 1993; Ostrom, 2005).

Policy process research follows diverse approaches each with a different investigative focus because, as Huntjens et al. (2011, p. 149) state, "policies change in a variety of different ways". John (2003) and Cairney (2013) give us insights into the complexity of the policy process. They point to five variables: (1) institutions that structure or influence behaviour, (2) policy networks of the actors, (3) exogenous factors, (4) the choices of actors, and (5) the ideas that help guide actors' behaviour. Policy actors include politicians, public officials, company managers, interest/advocate group members, academics, researchers, and active citizens (Moyson et al., 2017). The definition of policy process by Weible (2014) also indicates similar components: "the study of the interactions over time between public policy and its surrounding actors, events, and contexts, as well as the policy outcomes" (p. 5). A study on policy process may focus on a specific variable or investigate how variables of a policy process interact with each other.

The field of policy process analysis has gone through changes and diversified into different schools of thought over time. In the 1960s, the focus was on the policymaking process emphasizing policy outputs in legislative institutions (Weible, 2014). Another core issue of early studies on public policy was to divide the process stages. Lasswell's (1956) seven stages of the decision process are among the most influential models used in policy process analysis. The stages include intelligence, recommendation, prescription, invocation, application, appraisal, and termination. Jones (1970) developed Lasswell's categories into 'policy cycles' which consist of agenda setting, policy formulation, implementation, and evaluation. The policy cycle model is still widely used to explain the policy process (Weible, 2014; Howlett et al., 2009; Bridgman & Glyn, 2003).

Criticisms of linear approaches were made by Jenkins-Smith and Sabatier (1994), who pointed out that the policy cycle model lacks causal theory and is biased towards top-down processes. At the same time, new theories emerged with more focus on the causal explanations of policy change and the dynamics between various actors at different governance levels (Hogan & Howlett, 2015). These include the multiple stream approach (Kingdon, 1984), the institutional analysis and development framework (Kiser & Ostrom, 1982), the punctuated equilibrium theory (Baumgartner & Jones, 1991), the advocacy coalition framework (ACF) (Jenkins-Smith & Sabatier, 1994), the innovation and diffusion models (Berry & Berry, 1990), path dependency theory (Pierson, 2000), policy network analysis (Marsh & Rhodes, 1992), social constructionism (Schneider & Ingram, 1993), focusing events (Birkland, 1998), and policy entrepreneurship (Mintrom & Norman, 2009).

As with policy change, these theories have various foci and emphases. One of the most notable differences between theories of the policy process is whether agent or structure is given more weight. Human agents are the actors, whereas structure signifies the rules and resources of social systems (Giddens, 1984). The relationship between structure and agency is "one of the perennial issues or dilemmas of social science" (Hay & Wincott, 1998). Giddens (1984), who sees institutions as chronically reproduced rules and resources, states that understanding human agency and social institutions is one of "the tasks of social theory in general" (p. xvii).

Theories usually place particular emphasis on either agent or structure. For example, Kingdon's multiple stream approach focuses on the role of individuals (agents) who 'make things happen'. In this approach, individuals, timing, and context have more considerable influence than institutions over policymaking (Zahariadis, 2014). Kingdon (1984) argues that policy entrepreneurs exploit windows of opportunity, which open when problems, policies, and politics (the "three streams") meet as a result of a window of opportunity (Baumgartner & Jones, 2009; Zahariadis, 2014). The advocacy coalition framework (ACF) developed by Sabatier (1988) and Sabatier and Jenkins-Smith (1993, 1999) also underscores the importance of actors in policy processes. ACF explains that policy changes occur through policy learning by advocacy coalitions, although the explanation given on the role of ideas is ambiguous (Huntjens et al., 2011).

Meanwhile, new institutionalism approaches such as historical institutionalism and sociological institutionalism emerged through criticism of the tendency in social science to overemphasise agency without sufficient analysis of structure (i.e., rational-choice theory and behaviourism) (Schmidt, 2008). DiMaggio and Powell (1991) note that behaviourism theories prevailed in social science in the 1970s. Those theories, including public choice theory, valued the aggregate results of the rational choices made by individuals. However, the persistence of institutions cannot be reduced to the cumulative consequence of individuals' preferences. Furthermore, the power of institutions over individuals cannot be ignored as an aspect of social dynamics.

As research on the policy process cannot ignore either agent or structure, the two concepts are placed at opposite ends of a continuum, along which various theories can be placed (see Fig. 2.1). Between agent and structure, other vital components of the policy process such as ideas and norms can be located. Béland (2009) emphasises the role of ideas in the policy process, noting that although influential policy entrepreneurs are a key factor in successful policy ideas, the same ideas can empower traditionally weak actors under certain conditions. Civil rights and anti-racist ideas which helped Black women to mobilise against racial discrimination in the United States during the 1960s are good examples of this (Nadasen, 2005). Meanwhile, norms are shared interpretations of the world and reflect "unspoken premises" (Katzenstein, 1993, p. 268). Katzenstein (1996) observes that "norms typically inform how political actors define what they want to accomplish" (p. ix). Although norms are not necessarily in the form of formal regulations, they are institutionalised in a political community or a society and widely accepted by the group of actors.

In addition, the concepts of discourse and institution can be placed along the continuum, embracing either agent and idea (discourse) or norm and structure (institution). Generally, discourse analysis seeks to understand the ways that ideas are produced and proliferated by agents (Huitema & Meijerink, 2009). Meanwhile, the definition of institutions

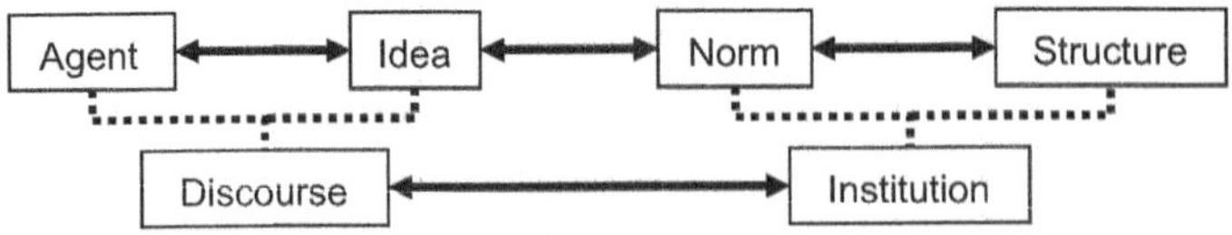

Fig. 2.1 The continuum of key components in policy process

embraces not just formal regulations and structures, but also informal norms and rules-in-use.

This study tries to investigate policy process from a balanced perspective between agent and structure. Those theories which put heavy emphasis on either agent or structure were not chosen for the analytical framework. As both agents and structures cannot be underestimated in policy process analysis, this study combines discourse and institutional analyses in order to see how both agent and structure affect the policy process. In the next section, the reasons why discursive institutionalism was chosen are explained in more detail.

2.1.2 Combined Framework: Discursive Institutionalism

There have been attempts to combine different approaches along the continuum for explaining policy change. The punctuated equilibrium theory demonstrates a compromised approach that incorporates both agents and institutions (Baumgartner et al., 2014). Founded in the concepts of political institutions and bounded rational decision-making, this theory assumes that policy is generally stable and pressure for change can be resisted. According to Baumgartner et al. (2014), the power, prestige, and legitimacy of policy monopoly are important elements of the policy process. Policy monopoly influences policymaking through powerful ideas or images which are usually connected to core political values. However, when the built-up pressure for change is sufficient, bursts of change or a massive intervention led by new actors can take place. Such large-scale changes are possible as public understandings of existing problems evolve. The new actors would restructure the rules and the balance of power for new institutions (Baumgartner et al., 2014).

While the punctuated equilibrium theory has significantly contributed to the policy process literature, some limitations for a balanced analysis of discourse and institution can be found. Firstly, it does not pay much attention to the role of discourse in the policy process. Nor does it discuss the subjectivity of information. Lastly, the power struggles for policy changes are not highlighted as much as the incremental changes during policy process (Baumgartner et al., 2014).

Discursive institutionalism attempts to better combine agent and structure frameworks with consideration of power relations. Schmidt (2010) provides a broad definition of discursive institutionalism. She sees discursive institutionalism as "an umbrella concept for the vast range of works in political science that take account of the substantive content of ideas and the interactive processes by which ideas are conveyed and exchanged through discourse" (Schmidt, 2010, p. 3).

When this broad definition of discursive institutionalism is applied, many theories related to ideas in the policy process can be included within its scope. Schmidt (2010) states that those who argue the importance of policy ideas (e.g., Kingdon, 1984), programmatic ideas or paradigms (Hall, 1993; Berman, 1998) or more profound philosophical ideas (Campbell, 2004) take a discursive institutionalism approach. Schmidt (2008) analyses scholars from the three different streams of new institutionalism (historical institutionalism, socio-cultural institutionalism, and rational choice institutionalism) who have taken a discursive institutionalism approach. The main reason why new institutionalists pay more attention to discourse is that they find the subordination of agency (action) to structure (rules) within institutionalism problematic (Schmidt, 2008).

Like other schools of institutionalism, discursive institutionalism acknowledges the significant role played by institutional circumstances in framing the discourse (Schmidt, 2009). These circumstances do not only include (relatively) static features of the society such as social norms and governmental systems but also refer to sudden changes such as economic crises, political unrest, natural disasters, and accidents. Discursive institutionalists also embrace historical institutionalist approaches to the examination of policy change and recognise that exogenous crises or shocks may influence change. They argue that it is necessary to look into both the institutional and historical contexts in order to understand the policy process.

The main feature of discursive institutionalism is that it explains policy change as well as policy continuity through the lens of discursive interaction between policy actors in an institutional context (Schmidt, 2010).

While institutions form the space in which discourses interact, discursive change may also lead to institutional change. This can be summarised by Giddens' (1990) explanation that institutions work as both "structures and constructs" of the discourse of political actors (as cited in Palmer, 2010, p. 997). When a discursive change occurs, it can lead to actions for institutional change individually or collectively (Schmidt, 2015). This is because when an alternative discourse appears, people are able to think outside their institutional frames and criticise the existing institutions. The emphasis on the role of discourses differentiates discursive institutionalism from other schools of new institutionalism.

Although interest in the role of ideas and discourse in the policy process has grown amongst scholars, there are no fixed methods for conducting discursive institutionalism research. This thesis proposes a new methodological approach to study discursive institutionalism. It first applies both a discourse and an institutional analysis to the case studies. In this way, the respective explanatory power of the two analytical approaches can be assessed (Parsons, 2007; Béland, 2009). As the next step, the interaction of discourses and institutions in each case is examined. On this basis, this study tries to explain the ways that policy either changes or remains stable.

Using analytical frameworks for discourse and institutions together allows the identification of causal mechanisms behind water management policy processes. By taking both discursive and institutional approaches, policy changes are explained not only with endogenous discursive interactions but also with exogenous processes such as shocking events (Schmidt, 2010). As Hall (2010) suggests, using multiple approaches can help find the points of 'tangency' between these approaches within policy process dynamics. Encouraging researchers to explore different approaches, Hall (2010) argues that "it is short-sighted to cling to one at the expense of benefiting from the others" (p. 27).

Working with more than one theory is a common practice in the field of public policy analysis. Cairney (2013) categorised multi-theoretical approaches in policy analysis in three ways. The first, "synthesis", is an attempt to generate a single theory by combining various theories, concepts, or models. The second, "complementary" approach, uses multiple approaches in order to explain an empirical phenomenon. Cairney

(2013) points out that in many cases research which makes a comparison of different explanations is superficial as it does not explore each theory thoroughly. The third approach is labelled as "the contradictory approach" by Cairney (2013, p. 3). It involves the comparison of multiple theories associated with respective empirical studies. In doing so, researchers may select or reject specific theories. Discursive institutionalism falls under the "synthesis" approach because it aims to explain policy processes by one theory which combines the concepts of discourse and institution.

2.1.3 Discursive Institutionalism Policy Process Analysis in Environmental Politics

Although 'discursive institutionalism' is relatively new, it has been used to study environmental politics based on a combined examination of ideas and institutions and their role as drivers of the policy process (Bosomworth, 2018; Arts & Buizer, 2009; Den Besten et al., 2014; Buijs et al., 2014; Palmer, 2010). This body of research concludes that discourse and institution are interdependent in environmental policy processes. Reviewing these previous studies helps us to gain insights into policy processes in the environmental sector. Relatively few studies exist making use of discursive institutionalism in relation to climate change adaptation or water management policy.

The Forest and Nature Conservation Policy Group at Wageningen University published a series of forestry policy analyses based on discursive institutionalism (Arts & Buizer, 2009; Den Besten et al., 2014; Buijs et al., 2014). Arts and Buizer (2009) analysed global forest politics from the 1980s to the 2000s. They find that new ideas and meanings of sustainability, biodiversity, and governance have been institutionalised in the forestry field, resulting in policy change. Den Besten et al. (2014) use a discursive institutionalism framework in their work on Reducing Emissions from Deforestation and Forest Degradation (REDD+), as part of the United Nations Framework Convention on Climate Change (UNFCCC). They conclude that the development of REDD+ between 2004 and 2011 in the international convention conferences illustrates how the production of ideas by actors and the institutionalisation of these ideas were mutually constitutive. The institutional contexts created the boundaries within which actors discussed new ideas on the scheme. Ideas and institutions are therefore "symbiotic and cannot exist separately" (Den Besten et al., 2014, p. 46). Buijs et al.'s (2014) study on the dramatic shift of Dutch nature conservation policy is also based on discursive institutionalism. They argue that critical discourses played an essential role in the sudden change of policy. Policy actors, including a then new state secretary, used existing counter-discourses to change nature policy and break down key institutional arrangements. Path-formation and path shaping occurred through discursive struggles and re-institutionalisation processes (Buijs et al., 2014).

Both Palmer (2010) and Kern et al. (2015) investigate the United Kingdom's energy policy change through a discursive-institutionalist perspective. Palmer (2010) concludes that ideas and institutions in the United Kingdom's biofuels policy interact with each other in a complex and interdependent way. Dominant ideas supported by the biofuel industry interact with institutions in a self-reinforcing manner, precluding significant changes in biofuel policy. He argues that this interplay was evident during the agenda-setting and policy change processes. Kern et al. (2015) explain UK energy policy change using Hall's (1993) policy paradigm analytical framework which includes an analysis of the power of ideas and institutional persistence. They observe that the UK energy policy between 2000 and 2011 was based on a pro-market energy policy paradigm, which aimed to "establish and maintain a competitive, freely trading energy market" (Kern et al., 2015, p. 274). It was believed that a free and competitive market would bring secure and affordable energy to the citizens. From the late 1990s, claims emerged that governments should take a more active role in energy policymaking in order to protect the climate. However, these claims only changed the policy paradigm to a limited extent, with the paradigm showing "quite high degrees of path dependency" (Kern et al., 2015, p. 277). What triggered policy change was the truckers' strike in 2000, an energy price rise, and Russia's gas transit dispute with Ukraine in 2008. Meanwhile, the climate narrative also gained global attention, and experts, think tanks, and NGOs called for the British government to show more government leadership in renewable energy development and deployment. The accumulation of impacts from these events and narrative changes resulted in a paradigm shift in the pro-market energy policy during the observed period.

Bosomworth's (2018) analysis of the fire management policy sector of Victoria, Australia, is another example of an adaptation policy analysis that takes a discursive institutionalism perspective. Bosomworth finds that current institutional logics in the sector restrain transformative governance for climate change adaptation. In the fire management sector, fire management is regarded as a technical problem and hazards are thought to be managed and controlled within the capacities of established organisations. Bosomworth (2018) argues that discursive institutionalism can identify the influence of discourse on institutions in various sectors related to adaptation so that the pathway to transformative governance can be found.

To conclude this section, discursive institutionalism offers abundant possibilities for utilising discourse and institutional frameworks for policy process analysis. Unlike previous studies, this research attempts to begin by applying discourse and institutional analyses separately for the case studies. After examining the discursive and institutional settings of each case, interactions between the two variables within policy process are investigated and compared between the two cases. To this end, the theoretical background and operational features of discourse and institutional analyses are presented in depth in the following sections.

2.2 Discourse Analysis

2.2.1 Theoretical Background of Discourse Analysis in Political Science

Like many other general or abstract concepts, discourse does not have a single definition. Fischer (2003) points out that authors in different academic traditions (i.e., hermeneutics, post-structuralism, and post-Marxism) use the term in different ways. As this book focuses on environmental politics, it is important to see how discourses are conceptualised by notable scholars in environmental discourse analysis. Hajer

(1995, p. 60) defines discourse as "a specific ensemble of ideas, concepts and categorisations". These ensembles are produced, reproduced, and transformed in society, creating as well as reflecting physical and social realities. Another definition is suggested by Dryzek (2013). He defines discourse as "a shared way of apprehending the world. ... it enables those who subscribe to it to interpret bits of information and put them together into coherent stories or accounts" (p. 9). Both scholars emphasise that discourses are embedded in the language (Hajer, 2006; Dryzek, 2013).

Discourse started to emerge as a major topic in political science in the 1980s after Michel Foucault's work, which addressed the relationship between power and knowledge, gained popularity in academia (Foucault, 1980). The concept rests on the assumption that language both shapes and reflects one's perspective of the world (Hajer, 2006). This assumption is rooted in the school of post-empiricism. Traditionally, social science has placed emphasis on causal explanations based on robust empirical evidence. In contrast, post-empiricists argue that explanations of society can be established in numerous ways (Fischer, 2003). This is not an attempt to underestimate the value of empirical research. Instead, it is an epistemological orientation that attempts to overcome the previously widespread view that "reality is objective" (Fischer, 2003, p. 13).

As the previous sentence indicates, post-empiricism has its foundation in social constructionism, which perceives social and political life as a creation of "social meanings produced and reproduced through discursive practices" (Fischer, 2003, p. 13). Berger and Luckmann (1967) wrote in their seminal book *The Social Construction of Reality*, "Society is a human project. Society is an objective reality" (p. 61). This means that there is no objective truth because facts are socially constructed by people based on their own values (Fischer, 2003). Socially constructed problems do not necessarily mean that the problems are not real (Dryzek, 2013). Social constructionists recognise subjectivity in presuppositions and assumptions, which affect decision-making processes (Berger & Luckmann, 1967; Fischer, 2003). The main focus of social constructionists' studies is not to find facts, but to find what the arguments about social problems are, how they are established and by whom (Best, 1989).

In the same vein, post-empiricists describe politics as a space in which actors try to create social meanings and fight to gain support for their subjective truth. This view is in line with many political scientists' arguments. Stone (2002) defines policymaking as "the struggle over ideas" (p. 11). Schön and Rein (1994) also state that actors amid policy controversies contest for defining the problematic situation. Often, the actors use strategies such as "naming and framing" of the policy situation to imply "not only what is at issue but what is to be done" (Schön & Rein, 1994, p. 29). Hajer's (1995) explanation about the nature of politics is in agreement with Schön and Rein and Stone. He states that "political conflict is hidden in the question of what definition is given to the problem, which aspects of social reality are included and which are left undiscussed" (Hajer, 1995, p. 43).

Policy analysis from a post-empiricist perspective means scrutinising the frames used in policy disputes. This is because not only the consequence of events or actions, but their interpretation can determine policy legitimacy (Fischer, 2003). This interpretation is based on beliefs and expectations, and consists of different thoughts about the way things are, what is to be done, by whom, and how to do it. In this way, discourses can steer power (Foucault, 1980). Consequently, the discourses which circulate those interpretations are the critical subject for a post-empiricist study.

2.2.2 The Elements of a Discourse Analysis

A discourse analysis typically starts by examining written or spoken statements. As mentioned above, this method is based on the assumption that language represents the socially constructed reality itself. At the same time, as Edelman (1988) explains, looking into the use of language provides a clue to the speaker's view. One can understand why and how certain actors in a society behave and respond in a specific way through comprehension of the language that the actors use.

The basis of discourse consists of assumptions, judgements, and contentions (Dryzek, 2013). In order to construct discourses, actors use frames, rhetoric, symbols, metaphors, narratives, and storylines (Hajer, 2006). In policy process studies and political science, analysis of these elements aims to expose the underlying assumptions, interests, values, beliefs, and concepts with which policies are drafted. These elements comprise the main focus of the discourse analysis conducted in this book.

The first element of discourse that will be investigated is frames. Indeed, given that discourse analysis takes into account the significance of frames in the policy process, discourse and frame analyses share commonality in several aspects. Frame analysts include discourse as part of their analysis and vice versa (Goffman, 1974; Schön & Rein, 1994; Hajer, 2006). The distinguishable point is that discourse analysis has its origin mainly in linguistics whereas frame analysis originated from communication and journalism studies. However, it can be assumed that the two analyses are not fundamentally different from each other when they are applied to policy process analysis. Although this book focuses on discourse analysis, some insights are adopted from frame analysis literature.

Schön and Rein (1994) define frames as the "underlying structures of beliefs, perception, and appreciation" (p. 23). Frames can be compared to a pair of coloured glasses through which one sees the world. Facts can appear different to individuals depending on which glasses one puts on. Socially constructing the problematic situation, a frame provides not only conceptual coherence, but also a direction for action and a basis for persuasion (Fischer, 2003). In this regard, a policy frame determines what the policy problem is, and consequently guides actions to be taken to solve the problem. In other words, a policy frame is "a normative-prescriptive story" (Rein & Laws, 1999, p. 93). Framing processes can legitimate certain discourses while marginalising others (Boykoff, 2011).

Policy controversies arise when 'the facts' are not agreed on by actors who frame the policy differently. Schön and Rein (1994) provide an example of this with their analysis of urban housing policy in the United States in the late 1950s and early 1960s. Regarding slums in large cities, at least two frames were present: one side argued that the slum areas were deteriorated and complete redesign was needed, whereas the other side argued that the communities in those areas are valuable for social organisation and integration. In such controversial situations, actors identify their interests in accordance with their frames. For instance, real estate developers and urban planners with the former frame base their interest in real estate profits, whereas the advocates for working-class neighbourhoods find preserving the neighbourhood valuable because they have the latter frame (Schön & Rein, 1994). Actors often choose specific evidence which can support their frames and dismiss other evidence. In this way, the frames can be reinforced. Schön and Rein (1994) point out that a frame can thereby hardly be disproved once it is adopted by actors.

The second element of discourse to analyse is storylines. A storyline is "a narrative on social reality" (Hajer, 1995, p. 62). Similar to frames, once a storyline is established, it does not easily break, as people interpret facts within their own storylines (Fischer, 2003). If the facts are not consistent with the storylines, it is not the storylines but the facts that are often denied or rejected. Douglas (2002) explains this in the context of pollution issues: the scientific evidence is less relevant to responses in society than the concept of pollution created by storylines. This example is in line with Fischer's (2003) argument that "politics has more in common with religion than with science" (p. 103). Fischer explains further:

> No amount of data, regardless of how well tested and verified it might be, will convince a person that anything important or useful has been presented if, in his or her view, the findings lead to policy judgements that take him or her in the wrong direction, or at least down a road he or she is unwilling to travel. (Fischer, 2003, p. 111)

Compelling stories can shape public consciousness about an issue at hand and guide the policy process, programme design, fund allocation, and evaluation (Schön & Rein, 1994). The influence of the new policy discourse rests on "the cognitive power of its storylines as well as on its attractiveness" (Hajer, 1995, p. 65). Fischer (2003, p. 167) argues that some stories are received as "more truthful and humane than others" and thus they have more political power. The aforementioned example provided by Schön and Rein (1994) indicates this. When the story of unsanitary and dark slum areas was influential in the United States in the 1950s, the public policy followed the idea of urban renewal. However, in the 1960s, the story of low-income but socially integrated communities gained support and this led to a negative reaction towards the ongoing urban renewal policies. In the later 1960s, the redevelopment policy in the Boston area was adjusted to include community participation (Schön & Rein, 1994).

According to social constructionism, it is often not fact or specific knowledge that influences an individual's political perspective, beliefs, or ideas. Rather, emotions or values can be greater factors in taking a particular stance. Powerful emotive responses can change policy direction significantly. That explains how and why narrative storylines influence politics. Fischer (2003) argues that individuals often identify themselves with a political stance without scrutinising the relevant facts. Furthermore, the political positions last, spread, and influence others through discursive reproductions and repetition (Fischer, 2003). Linguistic manipulations can therefore lead many people to change their view of reality.

McBeth et al. (2014) emphasise that compelling narratives have a higher level of influence than scientific or technical information over political actors and the public by presenting the Narrative Policy Framework for analysing policy process. This framework attempts to analyse policy narratives systematically in terms of their form and content and to see how certain narratives gain power (McBeth et al., 2014). The limitation of this framework is that it mainly attributes the policy output solely to narrative components. The power of the actors who use the narratives and the institutional structure which gives power to the actors are given little attention.

The final elements of discourse to be analysed are rhetoric and metaphor, which are efficient for discourse dissemination. The message that actors wish to deliver can be easily and quickly diffused by a word or a phrase if rhetoric and metaphor are appropriately used. Their use can persuade and convince (Best, 1989; Dryzek, 2013). Schön and Rein (1994) argue that metaphors usually lead to reactions based on ideas that are already established in our minds. Therefore, political actors attempt to use socially accepted idioms such as human rights, economic benefit, cultural traditions, or religious frameworks in stories to gain support for their discourses (Dryzek, 2013). The use of popular symbols in society could also be effective for this purpose. Well-known political leaders in history such as Franklin Roosevelt and Winston Churchill were also known as rhetoricians (McBeth et al., 2014). Using labels and symbols can strategically alter public opinion and most, if not all, policymakers are aware of this (Zahariadis, 2014). Therefore, rhetoric and metaphors can be regarded as practical "discursive resources" (Fischer, 2003, p. 80).

To understand a discourse, it is necessary to investigate not only the contents of a discourse, but also external elements such as the power relationship of the actors who construct and use the discourse and the institutional setting in which the discourses are made (Fischer, 2003). This relates to the 'agents' and the 'structures' at both ends of the continuum of key components in the policy process that was introduced in Sect. 2.1.1. As Best (1989) summarised, claims-makers and claims-making processes are as important subjects as the claims themselves for social constructivists' investigation. The analysis of these external elements of discourses requires looking into the "histories, roles, institutional contexts, and interests of those who advance them" (Fischer, 2003, p. 146). Below are the external elements of discourses which will be examined in the case study and cross-case comparison chapters (Chaps. 4–6).

First, power is a critical dimension when examining the external dimension of discourses (Hajer & Versteeg, 2005). Behind any prevalent discourse, there are powerful actors who can influence the determination of whose claim prevails. Fischer (2003) states that the nature and reproduction of the power structure within societal and political institutions need to be investigated in order to analyse discourses. It is universally observed that elites with political and economic power try to establish and cultivate hegemonic discourses beneficial to their own interests. In the case study chapters, actors with decision-making power are identified by assessing the policy processes within the respective governmental system.

If actors use common concepts, metaphors, and storylines, it can be inferred that the actors share a similar view of the world (Dryzek, 2013). They can be regarded as a group called a discourse-coalition or a community of meaning, which is the second external element of discourse that will be investigated in this research. In Hajer's (1993) words, a discourse-coalition is "the ensemble of a set of storylines, the actors that utter these storylines, and the practices that conform to these storylines, all organised around a discourse" (p. 47). New actors are easily attracted by well-established storylines or narratives, which are often characterised by a certain degree of ambiguity or multiple interpretabilities, and join a coalition (Huitema & Meijerink, 2009).

If different discourse-coalitions have contrasting discourses, they are likely to have conflicts with each other. In fact, there is almost always some degree of conflict over the meaning of political language. Fischer (2003) states the nature of political language clearly: "If there are no conflicts over meaning, the issue is not political, by definition" (p. 57). He further explains that there will be conflicting opinions on the causes, the intentions behind, and the expected consequences of an action or a statement, even if the actors agree on what to do. Therefore, public policy processes inherently include disputes (Schön & Rein, 1994). Discourse-coalition analysis aims to find out who forms coalitions with whom and to reveal where conflicts occur between coalitions.

The last external element to delve into deeply is the link between discourse and institutional practices. This element is connected to the institutional analysis conducted in parallel with discourse analysis in this study. Schön and Rein (1994) argue that discourses among actors are always spoken or written within an institutional context. This institutional condition may change, channel, strengthen, or diminish discourses. Particularly, discourses supported by formal or socially established institutions are more powerful than those without such support. For example, if a government consistently disseminates a specific discourse via regulations or media (e.g., propaganda) and blocks other discourses, the discourse has a better chance of being delivered to and accepted by the wider public.

Discourse institutionalisation does not only mean the founding of formal rules and regulations based on a particular discourse. If many people subscribe to a certain discourse when conceptualising the world, the discourse can be regarded as institutionalised within society (Hajer, 1993). In this case, even political leaders who do not believe in the discourse are constrained by it (Goldstein & Keohane, 1993). However, discourses are not permanent. They undergo a lifecycle in which they develop, crystallise, bifurcate, and dissolve (Dryzek, 2013). Nevertheless, institutionalisation makes certain discourses more resilient and long-lasting than others (Hajer, 2006). Figure 2.2 shows both internal and external elements of the discourse analysis.

Two central questions of discourse analysis are: "How is the policy issue being conceptualized or 'framed' by the actors to the debate? How

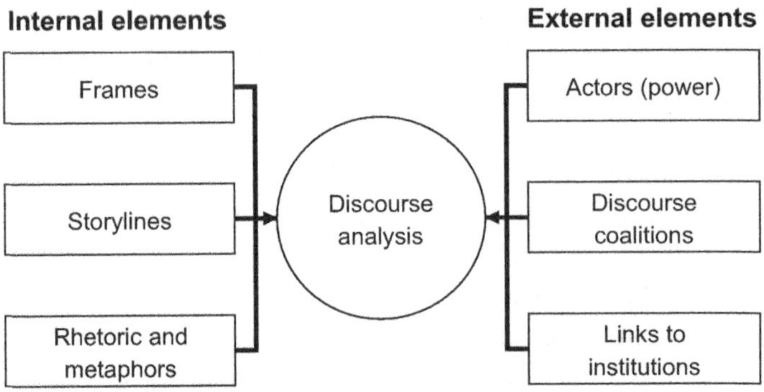

Fig. 2.2 The elements of the discourse analysis conducted for this book

is the issue selected, organized, and interpreted to make sense of a complex reality?" (Fischer, 2003, p. 143). These questions are guiding this research.

2.2.3 Significance of Discourses in Environmental Politics

Discourse analysis appeals especially in environmental policy and politics, where interactions among actors have been complex, generating various discourses (Hajer & Versteeg, 2005). Political actors frequently make coalitions or collide with each other based on these discourses. Several studies have been conducted on discourses in the environmental policy field (Hajer, 1995; Hannigan, 1995; Fischer, 2003; Litfin, 1994; Epstein, 2008; Dryzek, 2013). The subject varies from acid rain in Europe (Hajer, 1995) and ozone depletion (Litfin, 1994) to global environmental politics (Dryzek, 2013). Section 2.1.3 reviewed studies conducted on environmental policies through the lens of discursive institutionalism. In this section, the relevance of discourse in environmental politics is discussed further along with an overview of the current literature on this subject.

The high level of ambiguity and uncertainty around environmental problems is one of the reasons why discourse analysis is frequently used in environmental politics. Because environmental problems often stem from complex scientific phenomena, it is hard to rest on a simple causal relationship (Hajer, 2006; Wurzel, 2002). In uncertain situations, actors rely on the worldview with which they are familiar. Therefore, ideas serve as guides to behaviour (Fischer, 2003; Goldstein & Keohane, 1993; Katzenstein, 1993). With the increasing uncertainty surrounding climate change, discourse is gaining more attention.

People can interpret the same data in different ways and easily establish contrasting arguments in the name of the 'facts' (Hannigan, 1995). Hajer (1995) shows that different understandings of acid rain have generated different politics in respective European countries. When several arguments compete with each other, attractive narratives, rhetoric, and metaphors have the potential to convince actors in the policy process. This explains why, in environmental policy, debates over social and cultural factors can exert greater influence over the public view than any scientific 'facts' referred to by different actors (Fischer, 2003).

The history of environmental politics shows well that discourse is an essential variable in the shaping of environmental politics. Although environmental problems were gradually worsening throughout the twentieth century, the public did not react to the deteriorating environment until the 1960s (Hannigan, 1995). Thus, Hannigan (1995) points out that the public reaction to the environment is not automatic and, in many cases, public concern is independent of the severity of actual problems.

Most contemporary environmental issues such as acid rain, global warming, ozone depletion, toxic contamination, and radioactivity, are not visible to the public except in extreme cases. Therefore, the view of scientists, environmentalists, and media has had a great influence over discourse construction. Rachel Carson's book *Silent Spring* (1962) exemplifies how scientists' discourse, when supported by a convincing narrative, can affect public opinion and policy. Less than ten years after its publication, Earth Day 1970 was held and gathered 20 million people across the United States (Schreurs & Papadakis, 2020; Earth Day Network, 2020). The Environmental Protection Agency was created later in the same year.

Rhetorical tactics and strategies have been continuously used by environmental groups, experts, and their counterparts (Hannigan, 1995). Verbal and visual imagery is an effective way to catch attention. One of the most well-known examples of rhetoric about an environmental problem is the ozone hole. Although a hole does not exist in the ozone layer, the word 'ozone hole' was coined and a graphic image of a hole over the Antarctic was made to explain the thinning in ozone concentration. Subsequently, the image of the hole gained significant attention from the public as well as policymakers (Hannigan, 1995). Political leaders reacted to the growing concern worldwide and eventually formed the Montreal Protocol, an international treaty to phase out the chemical substances responsible for ozone depletion. Similarly, when the German environmental movement used the rhetorical term *Waldsterben* (translated from German as 'forest die-back'), the issue of acid rain attracted public attention (Hannigan, 1995). Opponents of the environmental movement also actively utilise rhetoric. The industry which produces genetically modified organisms, for instance, made quite a successful link between the technology, poverty reduction, and the fight against hunger (Hannigan, 1995).

Although post-modernists in environmental politics argue that terms such as 'nature' or 'climate change' are socially constructed, they do not deny the reality of environmental problems (Cronon, 1996; Dryzek, 2013). Dryzek (2013) notes that an extreme postmodern position has been used by climate change deniers (in the United States in particular) and that this is an incorrect adoption of social constructionism. He rightly asserts that "just because something is socially interpreted does not mean it is unreal" (Dryzek, 2013, p. 12). Hannigan (1995) points out that together with the mass media, science plays a role as a major institution in building environmental risks, knowledge, crises, and solutions of a society.

2.3 Institutional Analysis

2.3.1 What Institutions Mean

The second analytical framework adopted for this research is that of institutional analysis. Understanding the existing institutional context is a necessary step when analysing policies (Polski & Ostrom, 1999). Institutions have therefore been at the centre of political scientists' interest. Although the extent to which emphasis is attributed to the influence of institutions over the political process differs among scholars, they agree that the importance of institutional studies is not questionable.

The term 'institution' is used in various ways across disciplines. However, there is a common ground for the term agreed by most scholars. First of all, institutions are the result of human activities but they are not necessarily the products of purposive actions (DiMaggio & Powell, 1991). In this book, institutions do not mean organisations or physical structures. Instead, they refer to rules, including codified regulations or laws (formal) and social or cultural norms (informal) (Pahl-Wostl, 2015; Ostrom, 2011; Helms, 2014). Persistence is usually regarded as a common characteristic of institutions (Barkanov, 2016; Hall & Taylor, 1996). Although a group of institutionalists who contributed to the book *Explaining Institutional Change* (2010) edited by Mahoney and Thelen point out that gradual and incremental changes occur within institutions, those kinds of changes tend to be slow enough to be invisible to the actors. Another characteristic of institutions is repetitiveness. Institutions do not refer a single action. When actors regularly enact, re-affirm, and re-produce certain behaviours, their actions create or contribute to formalising an institution of the society (Beunen & Patterson, 2019). Lastly, members of a society tend to conform with institutions because institutions can provide incentives or disincentives (North, 1991). In this way, institutions provide stability to social life (Scott, 2008).

In political science, institutions used to mean formal laws, policies, rules, and organisations (March & Olsen, 1983). However, in the late 1970s new approaches that combined institutional analysis with contemporary theories emerged under the banner, 'new institutionalism'. New institutionalism emphasises the importance of informal institutions, which are usually invisible but shared by members of society (Polski & Ostrom, 1999; Ostrom, 2009). Rejecting the view that rational-actor models are the only possible explanations, new institutionalism recognises the cognitive and cultural background of actions (DiMaggio & Powell, 1991). This cultural and cognitive background is commonly understood as rules-in-use, norms, customs, and habits which are shared by more than two individuals, and recognised in repeated situations (Polski & Ostrom, 1999). New institutionalists delve into such informal institutions that individuals, groups, and organisations follow with the expectation of certain incentives.

Furthermore, according to new institutionalist perspectives, political institutions are not mere reflections of society, but also have power over society as political factors (March & Olsen, 1983). Therefore, institutions can become independent variables which can influence political actions. DiMaggio and Powell (1991) state that new institutionalism represents "an interest in properties of supra-individual units of analysis that cannot be reduced to aggregations or direct consequences of individuals' attributes or motive" (p. 8).

This study takes a new institutionalist approach which deals with not only formal institutions, but also informal institutions including norms and belief systems, and the dynamics of both in the policy process. Empirical studies since the 1990s have continuously shown that informal institutions play a role as an independent variable in policy processes (Katzenstein, 1993; Pahl-Wostl, 2015). However, under direct and indirect influence from informal institutions, formal institutions also exert power over politics. Therefore, formal institutions such as regulations and governmental structures (e.g., decision-making authorities and organizations in charge of certain policies) are also taken into account in this research.

2.3.2 Four Approaches in New Institutionalism

There are four different main approaches in the new institutionalist school within the field of political science, namely rational choice, sociological, historical, and discursive institutionalism (Schmidt, 2009; Hall & Taylor, 1996; DiMaggio & Powell, 1991). This study makes use of discursive institutionalism but it borrows some analytical elements from historical institutionalism for its analytical framework. This is because discursive institutionalism is an umbrella concept for the approaches that take account of ideas and the interactions within new institutionalism. Also, because discursive institutionalism does not have an agreed analytical method, it can be combined with other schools of institutionalism (Schmidt, 2010). This section introduces the distinctive features of the respective approaches before Sect. 2.3.3 presents the reasons why discursive institutionalism and historical institutionalism were used for this book.

Rational choice institutionalism is a calculus approach to institutions (Hall & Taylor, 1996). This approach considers that institutions are established because they are efficient for human society. This view regards individuals as interest maximisers and therefore, the institutions are 'self-reinforcing' (Mahoney & Thelen, 2010). Institutions reflect 'Pareto-optimal equilibriums' that change only when external shocks affect the system (Hall, 2010, p. 205). Institutions provide certainty and information to actors in uncertain and complex situations through a set of incentives and disincentives (DiMaggio & Powell, 1991). Furthermore, institutions are said to minimise transaction costs when actors follow them. In rational choice institutionalism, therefore, institutions are seen as the outcome of deliberate actions, rather than the unintentional results of human activities (DiMaggio & Powell, 1991).

Sociological institutionalism is also called 'sociocultural institutionalism' and takes a social and cultural approach to institutions (Hall & Taylor, 1996). It has the broadest definition of institutions among the four new institutionalist schools. Institutions are regarded as social patterns or codes of appropriateness and they are not necessarily the products of conscious design (DiMaggio & Powell, 1991; Mahoney & Thelen, 2010). In this view, institutions can be found everywhere in social life—from handshakes to strategic-planning departments (DiMaggio & Powell, 1991). Institutions are followed by actors not because they are efficient, but because it is appropriate to follow social norms and customs (DiMaggio & Powell, 1991). Individuals are assumed to be satisfiers who conform to norms. Sociologists argue that institutions are reproduced because there are no other conceivable alternatives for most individuals (Mahoney & Thelen, 2010). DiMaggio and Powell (1991) put it as "some of the most important sunk costs are cognitive" (p. 11).

Historical institutionalism uses a 'mixture' of the calculus and socio-cultural approaches. Historical institutionalists acknowledge both the cultural and calculus elements of institutions. They particularly trace 'processes over time' in terms of institutional stability and change. The longer institutions exist, the stronger the power of institutions to resist change. As Mahoney and Thelen (2010) explain, institutions are regarded within the school of historical institutionalism as "the political legacies of concrete historical struggles" (p. 7). This statement indicates that historical institutionalism recognises the role of power in building and maintaining institutions.

Pierson and Skocpol (2002) state that historical institutionalism has three important features in contemporary political science. Foremost, historical institutionalists tend to address big and substantive questions. Time is also strongly taken into account and the process tracing method is often conducted for the research. Therefore, the analysis highlights long-term processes, not only a slice of time. Finally, the main focus of the analysis is on the macro contexts and the configurations of institutions and processes. Through this, historical institutionalists attempt to understand the overall context of policy processes as well as the interactions within these processes (Pierson & Skocpol, 2002).

Finally, discursive institutionalism is a relatively new branch of the new institutionalism group, with an emphasis on the role of ideas and discourse in institutional change (Schmidt, 2009). Ideational institutionalism and constructivist institutionalism signify similar approaches to discursive institutionalism (Hay, 2006). As stated earlier in this section, discursive institutionalism is a rather broad term that refers to a range of studies on the relationship between ideas and institutions. The assumption behind this approach is that actors follow a logic of communication. Discursive interactions and their role in institutional changes are the main focus of discursive institutionalism (Schmidt, 2009).

Schmidt (2009) states that discursive institutionalism takes a very different approach to political reality than the other schools of new institutionalism. Discursive institutionalists regard institutions not just as a 'given condition' but also as constructs created by actors (Schmidt, 2008). In this context, discursive interactions may affect political reality by giving actors discursive power to overcome institutional constraints. One of the emphases discursive institutionalists make is "the use of ideas in the mass process of public persuasion" (Schmidt, 2009, p. 135). This is vis-à-vis "who talks to whom about what, when, how and why"—with institutions defining "where" these communications occur (Schmidt, 2009, p. 136).

As discursive institutionalism does not reject the central, older new institutionalist argument that institutions restrain actors, it can be used as a complementary explanation of political reality along with the other approaches. Indeed, with the 'turn to ideas' in political science, a number of new institutionalists from within the schools of rational choice, historical and sociological institutionalism, have employed discourse and ideas in their explanations. Schmidt (2008) asserts, therefore, that the boundary between discursive institutionalism and other new institutionalisms is sometimes fuzzy, and the institutionalists who accept the role of ideas and discourses in institutional change can be regarded as discursive institutionalists even though they do not label themselves as such. Particularly, historical institutionalism has been more accepting of the role of ideas in institutional changes than other new institutionalism schools (Schmidt, 2010).

There have been studies which use historical institutionalism alongside discourse and ideas. Schmidt (2008, 2010) refers to King (1999) and Weir (2006) as two historical institutionalists who have moved towards a discursive institutionalism approach within historical institutionalism. King (1999) analysed the importance of ideas and knowledge in the policymaking process of illiberal immigration policy in the United States and Britain, while also considering the institutional context of each country; whilst Weir (2006) argued that organised labour's effort to define itself as a political actor in the United States could be understood as an attempt to form a discourse-coalition. Their studies show that it is possible to combine discursive institutionalism with historical institutionalism for analysing policy process.

2.3.3 Discursive Institutionalism in Historical Institutionalism Tradition

Discursive institutionalism was chosen as the main analytical framework of this study. In addition, the analytical elements of historical institutionalism are used in the policy process analysis. The design of such an analytical framework is based on both the theoretical orientation and the empirical evidence of this study. This research follows the basic assumption of historical institutionalism institutionalists, who take a 'mixture' of the calculus and sociocultural approaches to institutions (Hall & Taylor, 1996). In addition, during the interviews and data analysis, it was found that institutional legacy and ideas have a great influence over environmental policies in both South Korea and Germany. For this reason, historical institutionalism was chosen over other new institutionalist approaches.

As explained in Sect. 2.1.3, discursive institutionalism emphasises the role of discourse more than other streams of new institutionalism including historical institutionalism. While there are historical institutionalists who have already paid attention to ideas (Hall, 1993; Berman, 1998), discursive institutionalists in the historical institutionalism tradition differentiate themselves by focusing on discourse as the endogenous factor of institutional change (Schmidt, 2010). The turn to the role of actors in the explanation of institutional change (and stability) makes the analysis of policy process more dynamic. In this regard, it was decided that discursive institutionalism has the most explanatory power for this book.

This study undertakes the discursive institutionalist analysis in three steps. First, social, programmatic, and policy discourses related to water management and climate change adaptation are investigated in each case. Second, it analyses the cases from a historical institutionalist perspective by searching for path dependency and critical junctures in each case. Third, the dynamic interactions between discourse and institutions are examined and compared between the two cases. In other words, the impact of discourses on institutional features and the ways that institutions influence discursive change are scrutinised. The assumption of this

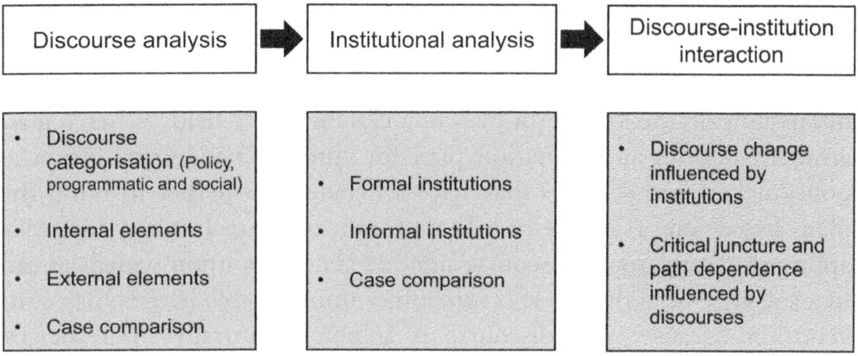

Fig. 2.3 Analytical framework flowchart

study is that by looking into the two-way dynamic of discourse and institution, the key factors underlying policy change can be understood.

Figure 2.3 shows the analytical framework flowchart of this study. The first step of the analysis will be to find different levels of discourses related to water management and climate change adaptation according to their scope, namely: policy, programmatic, and social discourses. This is a modified categorisation of Schmidt's (2008) analytical model, which defines ideas in political science. Schmidt (2008) argues that ideas exist on three levels: policy, programmatic, and philosophical. Policy level ideas indicate specific policies or policy solutions. The second level—programmatic ideas—encompasses more general programmes that function as the basis of policy. Programmatic ideas define the problems to be solved, the issues to be considered and the instruments to be applied (Schmidt, 2008). The third level consists of philosophical ideas, which are broader than the other two types of ideas. These are referred to as worldviews or "public philosophies" (Schmidt, 2008). While policy and programmatic ideas tend to be discussed regularly in the policy process, philosophical ideas are rarely exposed in the political scene except in times of crisis (Schmidt, 2008).

This book tries to classify discourses with a similar typology to Schmidt (2008). Policy discourse refers to the specific ideas and the surrounding conditions related to a particular problem. For example, solutions for flood risk management, such as arguments for building an embankment or establishing an alert system suggested by certain actors, are considered policy discourse. Meanwhile, programmatic discourse appears in the broader scope of policy design. Programmatic discourse can appear in discussions on the long-term plan of a certain policy field. When a government decides an adaptation plan for climate change, for instance, policymakers and scholars may have a debate on whether to build the plan based on a nature-based approach or a technology-intensive approach. Lastly, social discourse signifies general opinion widely spread in a society. Social discourse is sometimes unnoticeable in everyday conversation because such discourse is accepted naturally in a society. Nevertheless, social discourse has an impact on the policy process. Among the three types of discourse, social discourse is closest to the concept of norms because it is shared by a large number of people in a society.

What is different from Schmidt's categorisation of ideas is that this study takes into account the external aspects of discourse in the analysis. As Schmidt (2008) states, "discourse is a more versatile and overarching concept than ideas" (p. 309). As stated in Sect. 2.2.2, by using the term discourse, it is possible to indicate not only the ideas in the discourse, but also the context in which the discourse is created, conveyed, and utilised by different agents within the policy process. In addition, as this study intends to emphasise the nature of general discourse which is commonly communicated within a specific society (i.e., South Korea and Germany in this book), the term 'social discourse' is used instead of 'philosophical ideas' throughout this book.

After discourses are classified into three levels, the elements of the discourses are examined. Discourses have internal elements (frames, storylines, rhetoric, and metaphors) and external elements (actors, discourse-coalitions, and links to institutions). The discourses of each case are analysed accordingly and then compared to each other. In addition, the interrelationship of the different levels of discourses is investigated.

The second step of the analysis is to examine the institutional features of the two cases. Hall and Taylor (1996) point out that institutional analysis attempts to delve into how institutions affect the behaviour of individuals. Hall's book *Governing the Economy: The Politics of State Intervention in Britain and France* (1986) shows that institutions and organisations influence not just government policies, but also the strategies and coalitions of actors in a policy process. To investigate how institutions influence policies and actors' behaviour is not an easy process. The complexity of institutional analysis results from the fact that the pattern of interactions is usually uncertain and ambiguous (Polski & Ostrom, 1999). Although persistency is the main characteristic of institutions, it does not necessarily mean that actions and feedbacks can be easily predicted.

The Institutional Analysis and Development (IAD) framework developed by Ostrom and her colleagues suggests the direction of institutional analysis. Their framework emerged from an attempt to provide a general framework for institutional analyses in various disciplines (Ostrom, 2009). IAD focuses on the identification of action situations, which is a conceptual space in which actors interact with each other (Ostrom, 2011). However, its assumptions depend mainly on rational choice institutionalism and sociological institutionalism, rather than historical institutionalism. Therefore, it can be said that the IAD framework is helpful in conducting an institutional analysis but for historical analysis, it is not enough.

For analysing the institutions of the two cases, this study first attempts to identify formal and informal institutions related to water management and climate change adaptation. Although this study follows new institutionalism which heavily weights the importance of informal institutions, formal institutions cannot be ignored in the process of policymaking and policy implementation. Therefore, both formal and informal institutions are examined in each case and then compared between the cases.

The last step of the analysis is the investigation of the dynamic interactions between discourse and institution in policy process. The two-way interactions are examined respectively. First, the influence of institutions on policy discourse is investigated. Then the discussion moves on to the role of discourse in institutional changes. Particularly, the historical institutionalist analysis approach is used at this stage.

A distinct feature of historical institutionalism is that the persistence of institutions is explained with the term 'path dependence' and institutional changes with the term 'critical junctures'. The analyses of path dependence and critical junctures are inextricable because one should explain why institutional changes do or do not happen in order to understand the factors regarding institutional stability (Hall, 2010).

Path dependence is a positive feedback process which stabilises the existing institutional settings (Pierson & Skocpol, 2002). Pierson and Skocpol (2002) point out that patterns of political mobilisation, institutional rules, and citizens' ways of thinking can be explained by path dependence, which is a self-reinforcing dynamic. Even though some institutions are not in correspondence with general social norms such as freedom or justice, the powerful inertia of sticky institutional settings can cause them to persist. The sequence of events also determines which institutions are kept. This distinctive characteristic of institutions is also called 'increasing returns' which suggests "earlier events matter much more than later ones" (Pierson, 2000, p. 253). Once institutions are settled, interest groups and preferences develop based on the institutions, often becoming difficult to change (Pierson & Skocpol, 2002). To put it another way, historical factors limit the range of options to choose (DiMaggio & Powell, 1991).

Pierson and Skocpol (2002) took the health care and pension systems in the United States as an example of path dependence. While health care was initially provided by the private sector with the government attempting to intervene later, the public system for pension was developed first, with private companies later taking part in the system as supplementary providers. Consequently, the two different institutions (health care and pension systems) were supported by considerably different interest groups and attempts to change the system (e.g., the Patient Protection and Affordable Care Act) have faced enormous resistance.

From a new institutionalist perspective, institutions are usually persistent even if they are imperfect. This is explained by sunk costs (rational choice institutionalism), vested interests (historical institutionalism), or the difficulty of conceiving alternatives (sociological institutionalism) (DiMaggio & Powell, 1991). Nevertheless, from time to time, institutional changes take place. Those changes are usually episodic and dramatic, resulting from external/exogenous forces (DiMaggio & Powell, 1991). Triggers for change could include accidents, natural disasters, revolution, regime change, and war (Huitema & Meijerink, 2009).

Within historical institutionalism, fundamental changes can occur at critical junctures, which are often understood as the windows of opportunity when "usual constraints on action" are temporarily lessened (Mahoney & Thelen, 2010, p. 7). Critical junctures are often referred to as the primary feature of historical institutionalism (Pierson & Skocpol, 2002; Capoccia & Kelemen, 2007). In environmental movements, for instance, some critical events played the role of a critical juncture in environmental policy change. Massive environmental accidents, such as the Chernobyl nuclear accident (Van der Heijden et al., 1992), the Three Mile Island accident (Walsh, 1981), the Fukushima Daiichi nuclear disaster (Freiner, 2014), the Bhopal chemical disaster (Bullard & Wright, 1987), Exxon Valdez oil spill (Metcalfe, 2014), and Sandoz chemical spill, raised concerns over the environment (UBA, 2011). Consequently, in many cases, policy changes followed at national as well as international levels.

The main criticism towards the path dependence and critical juncture theories is that they rely on exogenous shocks or environmental shifts for institutional change (Mahoney & Thelen, 2010). Indeed, the possibility of endogenous, incremental change is very limited in the major institutionalist approaches. Nevertheless, some historical institutionalists perceive incremental changes do originate from within institutions. For example, Mahoney and Thelen (2010, p. 3) argue that "path-dependent lock-in is a rare phenomenon" and it is a problem of all the major institutionalisms that they see institutions as persistent and hard to change. They see institutional change as a result of a shift in the balance of power. If 'institutional power holders' are divided, or disadvantaged subordinate groups gain momentum to organise themselves and ally with other groups enough to "break prevailing institutional arrangements", institutional changes can happen (Mahoney & Thelen, 2010, p. 10).

Although Mahoney and Thelen's (2010) criticism towards historical institutionalism is well-grounded, their argument lacks an explanation of the process of power shift which triggers institutional change. In other words, it is still difficult to understand why and how power holders can be divided, or the subordinate groups can gain momentum. The advantage of using discursive institutionalism is that the cause of power struggles among actors can be explained by the conflicts between discourse-coalitions. When the actors' assumptions, judgements, and contentions are analysed through their discourse, the causal relationship between policy change and stability can be more clearly demonstrated.

References

Arts, B., & Buizer, M. (2009). Forests, discourses, institutions: A discursive-institutional analysis of global forest governance. *Forest Policy and Economics, 11*, 340–347. https://doi.org/10.1016/j.forpol.2008.10.004

Barkanov, B. (2016). *Institutionalism*. Edited by Encyclopedia Britannica. Retrieved December 9, 2017, from https://www.britannica.com/topic/institutionalism

Baumgartner, F. R., & Jones, B. D. (1991). Agenda dynamics and policy subsystems. *The Journal of Politics, 53*(4), 1044–1074.

Baumgartner, F. R., & Jones, B. D. (2009). *Agendas and instability in American politics* (2nd ed.). University of Chicago Press.

Baumgartner, F. R., Jones, B. D., & Mortensen, P. B. (2014). Punctuated equilibrium theory: Explaining stability and change in public policymaking. In C. Weible & P. A. Sabatier (Eds.), *Theories of the policy process* (2nd ed.). Westview Press.

Béland, D. (2009). Ideas, institutions, and policy change. *Journal of European Public Policy, 16*(5), 701–718. https://doi.org/10.1080/13501760902983382

Berger, P., & Luckmann, T. (1967). *The social construction of reality: A treatise in the sociology of knowledge*. Doubleday.

Berman, S. (1998). *The social democratic moment. Ideas and politics in the making of interwar Europe*. Harvard University.

Berry, F. S., & Berry, W. D. (1990). State lottery adoptions as policy innovations: An event history analysis. *The American Political Science Review, 84*(2), 395–415. https://doi.org/10.2307/1963526

Best, J. (Ed.). (1989). *Images of issues. Typifying contemporary social problems*. Routledge.

Beunen, R., & Patterson, J. J. (2019). Analysing institutional change in environmental governance: Exploring the concept of 'institutional work'. *Journal of Environmental Planning and Management, 62*(1), 12–29. https://doi.org/10.1080/09640568.2016.1257423

Birkland, T. (1998). Focusing events, mobilization, and agenda setting. *Journal of Public Policy, 18*(1), 53–74.

Bosomworth, K. (2018). A discursive–institutional perspective on transformative governance: A case from a fire management policy sector. *Environmental Research Letters, 28*(6), 415–425. https://doi.org/10.1002/eet.1806
Boykoff, M. T. (2011). *Who speaks for the climate? Making sense of media reporting on climate change*. Cambridge University Press.
Bridgman, P., & Glyn, B. (2003). What use is a policy cycle? Plenty, if the aim is clear. *Australian Journal of Public Administration, 62*(3), 98–102. https://doi.org/10.1046/j.1467-8500.2003.00342.x
Buijs, A., Mattijssen, T., & Arts, B. (2014). "The man, the administration and the counter-discourse": An analysis of the sudden turn in Dutch nature conservation policy. *Land Use Policy, 38*, 676–684.
Bullard, R., & Wright, B. (1987). Environmentalism and the politics of equity: Emergent trends in the black community. *Mid-American Review of Sociology, 12*(2), 21–37.
Campbell, J. L. (2004). *Institutional change and globalization*. Princeton University Press.
Cairney, P. (2013). Standing on the shoulders of giants: How do we combine the insights of multiple theories in public policy studies? *Policy Studies Journal, 41*(1), 1–21.
Capoccia, G., & Kelemen, R. D. (2007). The study of critical junctures: Theory, narrative, and counterfactuals in historical institutionalism. *World Politics, 59*, 341–369.
Carson, R. (1962). *Silent spring. Houghton Mifflin. Published in 2000 with a new introduction by P. Matthiessen*. Folio Society.
Cronon, W. (1996). The trouble with wilderness: Or, getting Back to the wrong nature. *Environmental History, 1*(1), 7–28.
Den Besten, J. W., Arts, B., & Verkooijen, P. (2014). The evolution of REDD+: An analysis of discursive-institutional dynamics. *Environmental Science & Policy, 35*, 40–48. https://doi.org/10.1016/j.envsci.2013.03.009
DiMaggio, P., & Powell, W. W. (1991). Introduction. In W. P. Walter & P. DiMaggio (Eds.), *The new institutionalism in organizational analysis* (pp. 1–40). The University of Chicago Press.
Douglas, M. (2002). *Purity and danger. An analysis of concepts of pollution and taboo: With a new preface by the author*. Routledge.
Dryzek, J. S. (2013). *The politics of the earth. Environmental discourses*. Oxford University Press.
Dye, T. R. (1984). *Understanding public policy*. Prentice-Hall.

Earth Day Network. (2020). *The history of earth day*. Retrieved April 2, 2020, from https://www.earthday.org/history/

Edelman, M. (1988). *Constructing the political spectacle*. The University of Chicago.

Epstein, C. (2008). *The power of words in international relations*. MIT Press.

Fischer, F. (2003). *Reframing public policy. Discursive politics and deliberative practices*. Oxford University Press.

Foucault, M. (1980). *Power/knowledge: Selected interviews & other writings 1972–1977*. Pantheon Books.

Freiner, N. (2014). Mobilizing mothers: The Fukushima Daiichi nuclear catastrophe and environmental activism in Japan. *ASIANetwork Exchange: A Journal for Asian Studies in the Liberal Arts, 21*(1), 27–41. https://doi.org/10.16995/ane.37

Giddens, A. (1984). *The constitution of society: Outline of the theory of structuration*. Polity Press.

Giddens, A. (1990). *The consequences of modernity*. Polity Press.

Goffman, E. (1974). *Frame analysis*. An Essay on the Organization of Experience.

Goldstein, J. L., & Keohane, R. (Eds.). (1993). *Ideas and foreign policy. Beliefs, institutions, and political change*. 6th paperback printing. Cornell University Press.

Hajer, M. A. (1993). Discourse coalitions and the institutionalization of practice: The case of acid rain in Britain. In F. Fischer & J. Forester (Eds.), *The argumentative turn in policy analysis and planning* (pp. 43–76). UCL.

Hajer, M. A. (1995). *The politics of environmental discourse. Ecological modernization and the policy process*. Clarendon Press; Oxford University Press.

Hajer, M. A. (2006). Doing discourse analysis: Coalitions, practices, meaning. In M. van den Brink & T. Metze (Eds.), *Words matter in policy and planning. Discourse theory and method in the social sciences* (pp. 65–76). Koninklijk Nederlands Aardrijkskundig Genootschap; Netherlands Graduate School of Urban and Regional Research.

Hajer, M. A., & Versteeg, W. (2005). Performing governance through networks. *European Political Science, 4*(3), 340–347. https://doi.org/10.1057/palgrave.eps.2210034

Hall, P. A. (1986). *Governing the economy: The politics of state intervention in Britain and France*. Oxford University Press.

Hall, P. A. (1993). Policy paradigms, social learning, and the state: The case of economic policymaking in Britain. *Comparative Politics, 25*(3), 275–296.

Hall, P. A. (2010). Historical institutionalism in rationalist and sociological perspective. In J. Mahoney & K. A. Thelen (Eds.), *Explaining institutional change. Ambiguity, agency, and power* (pp. 204–224). Cambridge University Press.
Hall, P. A., & Taylor, R. C. R. (1996). Political science and the three new institutionalisms. *Political Studies XLIV*, 936–957.
Hannigan, J. A. (1995). *Environmental sociology: A social constructionist perspective* (2nd ed.). Routledge.
Hay, C. (2006). Constructivist Institutionalism. In R. A. W. Rhodes, S. A. Binder, & B. A. Rockman (Eds.), *Oxford handbook of political institutions* (pp. 56–73). Oxford University Press.
Hay, C., & Wincott, D. (1998). Structure, Agency and Historical Institutionalism. In *Political Studies* (XLVI) (pp. 951–957).
Helms, L. (2014). Institutional analysis. In R. A. W. Rhodes & P. Hart (Eds.), *The Oxford handbook of political leadership* (pp. 195–209). Oxford University Press.
Hogan, J., & Howlett, M. (2015). Reflections on our understanding of policy paradigms and policy change. In J. Hogan & M. Howlett (Eds.), *Policy paradigms in theory and practice. Discourses, ideas and anomalies in public policy dynamics* (pp. 3–18). Palgrave Macmillan.
Howlett, M., Ramesh, M., & Perl, A. (2009). *Studying public policy. Policy cycles and policy subsystems* (3rd ed.). OUP.
Huitema, D., & Meijerink, S. (2009). *Water policy entrepreneurs*. Edward Elgar.
Huntjens, P., Pahl-Wostl, C., Rihoux, B., Schlüter, M., Flachner, Z., Susana, N., et al. (2011). Adaptive water management and policy learning in a changing climate. A formal comparative analysis of eight water management regimes in Europe, Africa and Asia. *Environmental Policy and Governance, 21*(3), 145–163. https://doi.org/10.1002/eet.571
Jenkins-Smith, H., & Sabatier, P. (1994). Evaluating the advocacy coalition framework. *Journal of Public Policy, 14*(2), 175–203. Retrieved January 20, 2020, from www.jstor.org/stable/4007571
John, P. (2003). Is there life after policy streams, advocacy coalitions, and punctuations: Using evolutionary theory to explain policy change? *Policy Studies Journal, 31*(4), 481–498. https://doi.org/10.1111/1541-0072.00039
Jones, C. O. (1970). *An Introduction to the Study of Public Policy*. Duxbury Press.
Katzenstein, P. J. (1993). Coping with terrorism: Norms and internal security in Germany and Japan. In J. L. Goldstein & R. Keohane (Eds.), *Ideas and for-*

eign policy. Beliefs, institutions, and political change (pp. 265–295). Cornell University Press.

Katzenstein, P. J. (1996). *Cultural norms and national security. Police and military in postwar Japan*. Cornell University Press.

Kern, F., Kuzemko, C., & Mitchell, C. (2015). How and why do policy paradigms change; and does it matter? The case of UK energy policy. In J. Hogan & M. Howlett (Eds.), *Policy paradigms in theory and practice. Discourses, ideas and anomalies in public policy dynamics* (pp. 269–291). Palgrave Macmillan.

King, D. (1999). *The name of liberalism: Illiberal social policy in the USA and Britain*. Oxford University Press.

Kingdon, J. (1984). *Agendas*. Alternatives and Public Policies.

Kiser, L. L., & Ostrom, E. (1982). The three worlds of action: A Metatheoretical synthesis of institutional approaches. In E. Ostrom (Ed.), *Strategies of political inquiry* (pp. 179–222). SAGE.

Lasswell, H. D. (1956). *The decision process: Seven categories of functional analysis*. Bureau of Governmental Research, College of Business and Public Administration, University of Maryland.

Litfin, K. T. (1994). *Ozone discourses. Science and politics in global environmental cooperation*. Columbia University Press.

Lowi, T. J., & Ginsberg, B. (1996). *American government*. Norton.

Mahoney, J., & Thelen, K. A. (2010). A theory of gradual institutional change. In J. Mahoney & K. A. Thelen (Eds.), *Explaining institutional change. Ambiguity, agency, and power* (pp. 1–37). Cambridge University Press.

March, J. G., & Olsen, J. P. (1983). The new institutionalism. Organizational factors in political life. *The American Political Science Review, 78*(03), 734–749. https://doi.org/10.2307/1961840

Marsh, D., & Rhodes, R. A. W. (1992). *Policy networks in British government*. Clarendon Pr.

McBeth, M. K., Jones, M. D., & Shanahan, E. A. (2014). The narrative policy framework. In C. Weible & P. A. Sabatier (Eds.), *Theories of the policy process* (2nd ed., pp. 225–266). Westview Press.

Metcalfe, J. (2014). *The 1969 oil spill that launched the modern environmental movement. Bloomberg CityLab*. Retrieved December 23, 2021, from https://www.bloomberg.com/news/articles/2014-01-29/the-1969-oil-spill-that-launched-the-modern-environmental-movement

Mintrom, M., & Norman, P. (2009). Policy entrepreneurship and policy change. *Policy Studies Journal, 37*(4), 649–667. https://doi.org/10.1111/j.1541-0072.2009.00329.x

Moyson, S., Scholten, P., & Weible, C. M. (2017). Policy learning and policy change. Theorizing their relations from different perspectives. *Policy and Society, 36*(2), 161–177. https://doi.org/10.1080/14494035.2017.1331879

Nadasen, P. (2005). *Welfare warriors. The welfare rights movement in the United States.* Routledge.

North, D. C. (1991). Institutions. *The Journal of Economic Perspectives, 5*(1), 97–112.

Ostrom, E. (2005). *Understanding institutional diversity.* Princeton University Press.

Ostrom, E. (2009). A general framework for analyzing sustainability of social-ecological systems. *Science, 325*(5939), 419–422. https://doi.org/10.1126/science.1172133

Ostrom, E. (2011). Background on the institutional analysis and development framework. *Policy Studies Journal., 39*(1), 7–27.

Pahl-Wostl, C. (2015). *Water governance in the face of global change.* Springer International Publishing.

Palmer, J. (2010). Stopping the unstoppable? A discursive-institutionalist analysis of renewable transport fuel policy. *Environment and Planning C Government and Policy, 28*(6), 992–1010. https://doi.org/10.1068/c09206j

Parsons, C. (2007). *How to map arguments in political science.* Oxford University Press.

Pierson, P. (2000). Increasing returns, path dependence, and the study of politics. *The American Political Science Review, 94*(2), 251–267. https://doi.org/10.2307/2586011

Pierson, P., & Skocpol, T. (2002). Historical institutionalism in contemporary political science. In I. Katznelson & H. V. Milner (Eds.), *Political science* (pp. 693–721). American Political Science Association.

Polski, M. & Ostrom, E. (1999). *An institutional framework for policy analysis and design.* Retrieved December 3, 2019, from https://pdfs.semanticscholar.org/ec83/18779f3f04c6a88cb59cdb338d4d8cde3b85.pdf

Rein, M., & Laws, D. (1999). Controversy, reframing, and reflection. In W. Salet & A. Faludi (Eds.), *The revival of strategic spatial planning: Proceedings of the colloquium. (pp. 93–108).* Royal Netherlands Academy of Arts and Sciences.

Sabatier, P. A. (1988). An advocacy coalition framework of policy change and the role of policy-oriented learning therein. *Policy Sciences, 21*, 129–168.

Sabatier, P. A., & Jenkins-Smith, H. (Eds.). (1993). *Policy change and learning: An advocacy coalition approach.* Westview Press.

Sabatier, P. A., & Jenkins-Smith, H. (1999). The advocacy coalition framework: An assessment. In P. Sabatier (Ed.), *Theories of the policy process* (pp. 117–166). Westview Press.
Schmidt, V. A. (2008). Discursive institutionalism: The explanatory power of ideas and discourse. *Annual Review of Political Science, 11*(1), 303–326. https://doi.org/10.1146/annurev.polisci.11.060606.135342
Schmidt, V. A. (2009). Comparative institutional analysis. In T. Landman & N. Robinson (Eds.), *Sage handbook of comparative politics* (pp. 125–143). SAGE Publications.
Schmidt, V. A. (2010). Taking ideas and discourse seriously: Explaining change through discursive institutionalism as the fourth 'new institutionalism'. *European Political Science Review, 2*(01), 1. https://doi.org/10.1017/S175577390999021X
Schmidt, V. A. (2015). Discursive institutionalism: Understanding policy in context. In F. Fischer, D. Torgerson, A. Durnová, & M. Orsini (Eds.), *Handbook of critical policy studies* (pp. 171–189). Edward Elgar Publishing.
Schneider, A., & Ingram, H. (1993). Social construction of target populations: Implications for politics and policy. *The American Political Science Review, 87*(2), 334–347. https://doi.org/10.2307/2939044
Schön, D. A., & Rein, M. (1994). *Frame reflection*. Basic Books.
Schreurs, M., & Papadakis, E. (2020). *Historical dictionary of the green movement*. Rowman & Littlefield Publishers.
Scott, W. R. (2008). *Institutions and organizations: Ideas and interests* (3rd ed.). Sage Publications.
Stone, D. (2002). *Policy paradox*. W.W. Norton.
Umweltbundesamt (UBA). (2011). *Sandoz chemical spill 25 years on*. Retrieved December 23, 2021, from https://www.umweltbundesamt.de/en/press/pressinformation/sandoz-chemical-spill-25-years-on
Van der Heijden, A., Koopmans, R., & Giugni, M. (1992). The west European environmental movement. In M. Finger (Ed.), *The Green movement worldwide* (pp. 1–40). JAI Press.
Walsh, E. J. (1981). Resource mobilization and citizen protest in communities around three Mile Island. *Social Problems, 29*(1), 1–21. https://doi.org/10.2307/800074
Weible, C. M. (2014). Introducing the scope and focus of policy process research and theory. In C. Weible & P. A. Sabatier (Eds.), *Theories of the policy process* (2nd ed., pp. 3–23). Westview Press.

Weir, M. (2006, August/September 30/2). '*Labor identities and coalition building in the United States*'. Paper prepared for delivery at the Annual Meeting of the American Political Science Association.

Wurzel, R. (2002). *Environmental policy-making in Britain, Germany and the European Union. The Europeanisation of air and water pollution control.* Manchester University Press.

Zahariadis, N. (2014). Ambiguity and multiple streams. In C. Weible & P. A. Sabatier (Eds.), *Theories of the policy process* (Vol. 2, pp. 25–58). Westview Press.

3

The Linkage between Climate Change Adaptation and Water

3.1 Adaptation to Climate Change

3.1.1 The Impact of Climate Change on Social-Ecological Systems

Anthropogenic climate change is a scientifically confirmed phenomenon. The most recent report of the Intergovernmental Panel on Climate Change (IPCC) reaffirmed that human activities have caused global warming (IPCC, 2021). According to the IPCC, the global mean surface temperature started to increase in the late nineteenth century (IPCC, 2013) with the average temperature of land and ocean now already 0.99 degree Celsius (°C) warmer than the period 1850–1900, which is taken as an approximation of pre-industrial temperatures (IPCC, 2021). According to the World Meteorological Organization (WMO) (2021), the years 2015–2020 were the warmest since 1850.

The main drivers of climate change are the growing concentrations of greenhouse gases (GHG) in the atmosphere, which are now at their highest level in the last 22,000 years (IPCC, 2013). Greenhouse gases include water vapour (H_2O), carbon dioxide (CO_2), methane (CH_4), nitrous

Y. h. Kang, *Climate Change Adaptation in River Management*, Palgrave Studies in Water Governance: Policy and Practice,
https://doi.org/10.1007/978-3-031-10486-2_3

oxide (N_2O), ozone, chlorofluorocarbons (CFCs), and hydrofluorocarbons (HCFCs). Among those, CO_2 is discussed the most because approximately 78% of the total GHG emission increase between 1970 and 2010 was due to CO_2 emissions (IPCC, 2014b). Research has shown that cumulative CO_2 emissions and global mean temperatures are correlated (IPCC, 2021). The foremost source of CO_2 emissions is fossil fuels, which have become the primary energy source for global economic activities since the Industrial Revolution in the mid-eighteenth century. In addition, deforestation, urbanisation, and other land-use changes have contributed to increasing CO_2 emissions, although the extent of contributions varies by country.

As a consequence of climate change, extreme weather events such as heatwaves and heavy precipitation are occurring more frequently. Mean precipitation is 'very likely' to rise in high and mid-latitude regions in the future. In addition, heavy precipitation is expected to occur more intensively and more frequently over land as the water cycle changes (IPCC, 2021).

Temperature rise and the shift of precipitation patterns are affecting social-ecological systems (IPCC, 2014a). The number of scientific publications on the impact of climate change on different sectors and regions is increasing (IPCC, 2014a). The Millennium Ecosystem Assessment (2005) concluded that climate change is one of the most significant drivers of biodiversity loss. The term 'Anthropocene', which implies the massive scale of human impact on the environment, was proposed to describe the current epoch (Steffen et al., 2018; Crutzen, 2006). The concept suggests that human activity is largely responsible for current global environmental change and that this is leading us out of the previous geological epoch, the Holocene (Steffen et al., 2011).

As far as social systems are concerned, climate change has affected and will affect various sectors. Notably, extreme weather events such as floods, heat and cold waves, droughts, cyclones, and wildfires have major effects on social systems. Unexpected weather conditions put regular farming practices in peril and so agriculture could be affected. Accidents related to torrential rain, wind gusts, and landslides could also increase (IPCC, 2014a). According to one of the medium emissions scenarios projected by the IPCC, in the absence of adaptation measures, the percentage of the population affected by river floods could double in the EU by the 2080s as compared to the period of 1961–1990 (EEA, 2016). Consequently, human health, well-being, and livelihoods may be threatened by changes in the water cycle.

It is worth noting that variation in vulnerability to climate change impacts depends not just on ecological sensitivity, but also on social conditions (Lindseth, 2005). Although the meaning of vulnerability is generally contested, it can be defined as susceptibility to injury (Handmer & Dovers, 1996). High vulnerability means less resilience to changes or shocks. Socioeconomic conditions such as poverty, status, ethnicity, and age have significant influence on the degree of vulnerability to climate change impacts (IPCC, 2014a). For example, two-thirds of heat stroke-related deaths in Seoul during the record hot summer in 2018 were people who had lived in extreme poverty (Hwang & Hwanbong, 2018). This shows that the impact of climate change on societies is closely related to social equality and fairness.

3.1.2 What Is Adaptation?

Given that climate change is already occurring and will not be reversible within the foreseeable future, adaptation to climate change is vital and unavoidable (Huntjens et al., 2011). The term adaptation has been used in other disciplines such as ecology and psychology; it began to appear in climate change literature in the 1990s when the impact of climate change became more visible and the international discussion on climate change became more pronounced (Fünfgeld & McEvoy, 2014; Moser & Ekstrom, 2010). The IPCC AR5 takes a broad definition of adaptation: "the process of adjustment to actual or expected climate and its effects" (IPCC, 2014a, p. 40). This definition is broad and can be applied not only to human societies but also to ecosystems.

If the focus is narrowed to the adaptation of human societies, which is the focus of this study, adaptation can be defined as the reduction of vulnerability (Smit et al., 2000; Eriksen et al., 2011); however, this is still an expansive definition. There are different views on adaptation and diverse interpretations exist, and these ideas have changed over time (Boykoff et al., 2013). Pelling (2011) points out that as the number of studies on adaptation grows and diverse fields become involved in the topic, more interpretations of adaptation appear, reflecting different worldviews. Among others, this study refers to the definition given by Nelson et al. (2007). They define adaptation as "the decision-making process and the set of actions undertaken to maintain the capacity to deal with change" (Nelson et al., 2007, p. 397). This definition serves as the basis of political science analyses, particularly those that focus on the policy process, which is the subject of this book.

Pursuing adaptation does not necessarily mean giving up climate change mitigation efforts. Just as GHG emissions reduction is crucial to preventing further changes in the climate system, taking action on adaptation to climate change is also essential. This is because a large part of the ongoing environmental change is inevitable despite current efforts to stabilise GHG emissions. What is more, adaptation activities may enhance mitigation activities when the activities are carefully planned and designed to avoid trade-off effects with mitigation. Recognising the necessity of adaptation strategies, a number of interrelated initiatives are being developed across global, national, and sub-national levels (Smit et al., 2000; Ayers & Huq, 2009; UBA, 2016; Adaptation Fund, 2017). Furthermore, Pelling (2011) asserts that mitigation activities are acts of adaptation in view of reducing the impact of climate change on societies, although not every mitigation initiative will be beneficial for adaptation.

Adaptive responses to climate change can occur in many different ways and forms (Smit et al., 2000). Leary (1999) points out that adaptation is not limited to building physical infrastructures such as sea walls. Climate adaptation also refers to behavioural adjustments that different actors make at different levels. Individuals and households may change their practices according to the changing environment. Formal and informal institutions such as procedures, legislation, or regulations may also be amended in order to reduce vulnerability (Leary, 1999; Eriksen et al., 2011). The discussions as to what adaptation activities include have been influenced by major actors on the "stage of climate science policy", namely climate scientists, politicians, environmental activists, and industry representatives (Boykoff et al., 2013, p. 241).

Reflecting the diversity in types of adaptation, there have been attempts to establish a typology of adaptation. The first attempt to categorise adaptation types was by Carter et al. (1994), who distinguish autonomous adjustments from planned adaptation. Bijlsma et al. (1996) see the functions of actions such as retreat, accommodate, or protect as the criteria for categorising adaptation. Biagini et al. (2014) identify ten types of adaptation activities: capacity building, management/planning, practice/behaviour, policy reform, information, physical infrastructure, warning/observing systems, green infrastructure, financing, and technology.

This study presents a typology suggested by Smit et al. (2000) with some adjustments: spatial scope (local, national, supra-national, and global levels), timing relative to stimulus (anticipatory, reactive), intent (autonomous, planned), and degree of change (resistance, transitional, transformational) (see Fig. 3.1). Smit et al. (2000) mention forms of adaptation (technological, behavioural, financial, institutional, or informational) as another category, but this book discusses different forms in the adaptation discourse section (Sect. 3.1.4) because forms of adaptation are related to problem framing.

The significant difference in this typology as compared to previous analyses is that the types are in a continuum in some of the areas (space, timing, intent, and the degree of change). This is because many adaptation actions hold multiple characteristics across the scale or types and cannot be categorised into a single type. Activities by different actors can also be interconnected with each other so it is hard to define the whole process as a single type. For instance, after a flood occurs, governments usually react to prevent the next flood by strengthening embankments or searching for an alternative solution such as recovering floodplains. This action can be regarded as both anticipatory and reactive. Building a knowledge-sharing network among local governments or non-governmental organisations in different countries is another example of an

Areas	Types of adaptation			
Space	Local	National	Supra-national	Global
Timing	Anticipatory			Reactive
Intent	Planned			Autonomous
Degree of change	Resistant		Transitional	Transformational

Fig. 3.1 A typology of adaptation activities

adaptation activity that cannot be categorised as simply a local, national, or global level action.

This study focused on certain adaptation types within the continuum, due to the limits of resources and time of the author. First, the cases are public policies at the *national level*. Except for emergency relief or restoration works after natural disasters, public policy for adaptation usually aims to prevent damage and manage anticipated risks due to climate change. Water management in particular can be categorised in preventive measures. With this in mind, this book deals with adaptation concerning *anticipatory and planned measures*. Lastly, this book uses *all the degrees of change* for comparing the two cases (see Sect. 3.1.2 for more details).

3.1.3 Institutional Development for Adaptation

Institutions are one of the crucial factors in determining adaptation (Adger, 2000; Olsthoorn & Tol, 2001; Naess et al., 2005), as institutions have a significant influence on "the social distribution of vulnerability" which leads to different degrees of adaptive capacity in society (Naess et al., 2005). Since climate change adaptation is a relatively new subject for policy, formal institutional settings are still new. In this section, international- and national-level institutions for adaptation are discussed in their historical context in order to understand the influence of international-level discussions on the policy process at the national level.

Climate change appeared in public debate in the late 1980s (Boykoff, 2011). The study on global mean temperature rise and CO_2 emissions conducted by James Hansen at the National Aeronautics and Space Administration (NASA) was reported as a front-page story in the *New York Times* in 1981. Gradually, international organisations began to establish climate policy. Primarily, the International Council of Scientific Unions, the United Nations Environment Programme (UNEP), and the WMO initiated relevant activities. The heatwave and drought in North America in 1988 were the main trigger for large-scale media attention to climate change (Boykoff et al., 2013). The UNEP and the WMO created the IPCC in the same year. The focus of the IPCC was mainly on understanding the global warming trend, the climate system dynamics behind the trend, the causes of climate change, greenhouse gas emission sources, and climate modelling. At the Rio Summit in 1992, the United Nations Framework Convention on Climate Change (UNFCCC) was adopted for more coordinated actions on climate change mitigation.

Today, the IPCC and the UNFCCC are the most influential international institutions in the direction of international climate change policy (Pelling, 2011). Since its establishment, the IPCC regularly releases Assessment Reports (ARs), which provide scientific evidence of climate change, its impacts, and mitigation options. The UNFCCC consists of 197 parties (as of 2021), which represent almost all the countries in the world. These two institutions work closely together. While the IPCC provides a scientific basis for the UNFCCC, funding and political will for IPCC research and activities are provided through the UNFCCC (Pelling, 2011).

Although the focus of the UNFCCC and IPCC at their outset was primarily on GHG emissions reduction (climate change mitigation), adaptation has gradually gained more importance within these institutions. The IPCC began publishing a separate report on climate change impacts, adaptation, and vulnerability in its Third Assessment Report in 2001 (McCarthy et al., 2001). Scientists began to recognise that mitigation efforts alone are not sufficient to address climate change impacts (Biesbroek et al., 2010). It only became a public issue in the media, however, when extreme weather events occurred in the United States and Europe. The heatwave in Europe in the summer of 2003 caused more than 35,000 deaths across Europe. Although it was debated whether the heatwave could solely be attributed to human-made climate change, this extreme weather became another trigger in the climate change discussion. This time, discussions on preparedness for climate change emerged. In the United States, Hurricane Katrina in August 2005 made adaptation discussions visible in the public arena. The hurricane hit the Gulf Coast and claimed more than 1800 lives (The Editors of Encyclopaedia Britannica, 2020). Boykoff et al. (2013) point out that after Hurricane Katrina, discussion on many adaptation-related issues such as risk, hazards, vulnerability, and responsibility expanded.

Discussion on adaptation in global media was intensified by the publication of *Stern Review: The Economics of Climate Change* (2006), which estimated the cost of climate change mitigation and impacts. As an economist, Nicholas Stern asserted that climate change could cost the world at least 5% of GDP per year and that it could increase to more than 20% of GDP per year if carbon emissions are not reduced (Stern, 2006). Furthermore, Al Gore's documentary film on climate change causes and consequences, *An Inconvenient Truth*, was released in the same year and attracted worldwide attention. Luke (2008) describes the film as a popular culture phenomenon that influenced the political debate about global warming. Scientists and researchers also continued suggesting the future effects of climate change and asserting the need for adaptation. The Fourth Assessment Report of the IPCC (2007) was another watershed in the adaptation policy agenda (Fünfgeld & McEvoy, 2014). The Assessment Report stated with very high confidence that human activities caused global warming and future climate change impacts could not be avoided (IPCC, 2007).

The necessity of adaptation actions has been increasingly recognised in the UNFCCC conferences. Following the Seventh Conference of the Parties (COP) held in Marrakesh in 2001, the UNFCCC has encouraged all its parties to identify priority actions and establish national adaptation plans. In the COP 16 in 2010, parties agreed to enact the establishment of a national adaptation plan (NAP) in all the member states. In the Paris Agreement adopted at COP 21 in 2015, adaptation is emphasised as equally as mitigation (United Nations, 2015).

Supranational adaptation policies are also initiated in Europe and among small island countries. In 2013, the European Commission adopted the EU Adaptation Strategy, which states support to its Member States in adaptation such as: providing guidance and funding, promoting knowledge generation and information-sharing, and enhancing the resilience of key vulnerable sectors (EEA, 2014). The EU has also set the goal of spending at least 20% of its budget from 2014 through 2020 for climate change-related action, including adaptation (European Commission, n.d.). Forty-four small island countries in Africa, the Caribbean, the Indian Ocean, the Mediterranean, the Pacific, and the South China Sea have formed the Alliance of Small Island States acknowledging their particular vulnerability towards sea-level rise and other adverse effects of climate change.

The national level is a key analytical level in adaptation because it lies between the international level (where adaptation is discussed) and the local level (where adaptation is in practice). National-level adaptation institutions vary by state. Many OECD countries adopted comprehensive multi-sectoral adaptation plans and strategies (Bauer et al., 2012; Biesbroek et al., 2010; Bauer & Steurer, 2015). In the national communications submitted to the UNFCCC, many governments report that national adaptation plans and strategies are established, and they assign adaptation works to their ministry of environment. Such an institutional setting for adaptation is in line with the fact that carbon emissions reduction tasks fall under the ministry of environment or natural resources (Connor & Rast, 2009).

Current adaptation plans and strategies at the national level submitted to the UNFCCC have several limitations. First, the concept of 'adaptation policy' is still not clearly defined and different understandings can be found among national governments (Dupuis & Biesbroek, 2013). Likewise, there have been policies on response to extreme weather events or to reduce disaster risks even before climate change adaptation became a policy agenda. However, because such policies are not explicitly called 'adaptation policies', their weaknesses or contribution to adaptation are not appropriately acknowledged (Füssel, 2007; Smit & Wandel, 2006; Dupuis & Biesbroek, 2013). Another challenge is that current national adaptation strategies in many countries are mainly functioning as a symbolic policy in domestic politics. Symbolic policies are useful to raise awareness about adaptation but concrete action plans are needed for solving policy problems (Dupuis & Biesbroek, 2013).

Because of such institutional features, mainstreaming of adaptation in other policy agendas such as economy, education, and finance is one of the most important issues concerning those who are tasked with adaptation in environmental ministries. The path-dependent nature of government institutions is a barrier to adaptation (Barnett et al., 2015). Institutions are usually resistant to change and their path-dependency could slow down changes necessary for adaptation (Barnett et al., 2015). Nevertheless, extreme weather events, the development of supra-national policies, the estimate of current and future damage costs, and relevant results from scientific research have provided the primary impetus for many national adaptation policies (EEA, 2014).

3.1.4 Adaptation Discourses (Programmatic Discourses)

This section reviews the literature on adaptation discourses and identifies three distinctive discourses in the adaptation field. This typology is utilised in Sect. 3.2.3 which deals with programmatic discourses on water management.

Actors in the adaptation policymaking process organise their ideas through specific frames. Policies are often planned and implemented accordingly (Fünfgeld & McEvoy, 2014; Juhola et al., 2011). Due to the uncertainty of climate change, discourses and the interpretations of scientific assessment often have a significant influence on adaptation policies. Therefore, it is necessary to pay attention to the underlying ideas and frames of adaptation policies (Fünfgeld & McEvoy, 2014; Tanner & Allouche, 2011; Eriksen et al., 2015). For example, the adaptation policy used in responding to sea-level rise can differ depending on the frames that actors use. If a "securing private property" frame dominates, measures can include strengthening a seawall for coastal defence. In contrast, if the adaptation discourse emphasises "nature conservation", rezoning of coastal areas would be chosen (Fünfgeld & McEvoy, 2014, p. 609).

Despite the importance of discourses within the adaptation policy field, the number of academic studies that have analysed adaptation policy discourses is not large (Juhola et al., 2011; Lindseth, 2005; Fünfgeld & McEvoy, 2014; O'Brien et al., 2007; Oppermann, 2011). Lindseth (2005) identifies three kinds of discursive strategies for climate change adaptation (scientific–economic, communicative–economic, and scientific–communicative discourse). Local institutions and actors can utilise those discourses as resources for adaptation policy planning. While Lindseth's strategy framework provides a suggestion of possible discourses for actors in the policy process, there are other frameworks that try to categorise existing discourses. Juhola et al. (2011) found four major frames for adaptation in Finland, Sweden, and the United Kingdom: existing measures, planning, vulnerability, and economic risk. The principal framings in those countries were normalised into European institutional policies and practices (Juhola et al., 2011). Meanwhile, Fünfgeld and McEvoy (2014) argue that adaptation is commonly framed within existing discourses such as disaster risk management and vulnerability reduction.

The explanatory gap of those studies is that they do not take into account the development and diversification of adaptation discourses that have occurred over time. Based on O'Brien et al.'s (2007) discussion on global environmental change discourses, this book suggests three categories for adaptation discourses within the historical context: (1) adaptation as technical-fix, (2) adaptation as socio-political task, and (3) adaptation as ecosystem service recognition. The main criterion used for the categorisation is the measures believed to be most effective for adaptation. The reason behind such a decision is that the frame of 'effective measures' for adaptation has changed over time, and also this criterion is a pragmatic lens for understanding adaptation policies implemented in practice.

The first discourse, '*adaptation as technical fix*', sees climate change as a threat to human society, and technical fixes through technological advancement as an important solution for fighting against nature's challenge. Adaptation was first considered to be a technical or individual matter of an apolitical problem (Eriksen et al., 2015; Andersson & Keskitalo, 2018). This is the most common approach around the world where adaptation is planned and implemented by governments (Resurreccion et al., 2008). For example, Oppermann (2011) points out that the UK Climate Impacts Programme's report, *Climate Adaptation: Risk, Uncertainty and Decision-Making* appears to problematise adaptation through a techno-scientific perspective with a focus on risk. This type of discourse was formed in the earlier stage of discussions on adaptation within the international policymaking scene. As indicated in the course of how the concept of adaptation was spread (Sect. 3.1.3), natural scientists were the main actors at the early stage of the discussion, and their work on adaptation was conducted with a view of adaptation as a technical agenda (Pelling, 2011). As mentioned by Eriksen et al. (2015), the IPCC reports have been criticised for relying heavily on global models produced by northern scientific institutions, while local studies on vulnerability and poverty are rarely conducted. Such perspectives may lead to solutions based on the application of technology.

The techno-centric bias of the IPCC was transferred to the UNFCCC and subsequently, to the national adaptation plans by national governments (Pelling, 2011). For example, Resurreccion et al. (2008, p. 18) note that the approach to adaptation in Southeast Asia has tended to be linear and straightforward and that outcomes have leaned towards "technical fixes" which were similar to mitigation policies. They further argue that Southeast Asian countries emphasise scientific methods such as modelling and scenario building because most of the actors involved in climate change policy process have natural science backgrounds (Resurreccion et al., 2008).

The techno-centric approach is also preferred by many actors in politics, business, and finance, as this kind of adaptation does not require restructuring of the economy or institutions (Leach, 2015). Technological advancement is widely believed to contribute to economic growth. This discourse can be found in arguments supporting green growth or large-scale projects such as hardening shorelines and new infrastructure construction (aquifer storage, on-site power supply, flood barriers, etc.) (Climate Change Adaptation Resource Center, 2019). Green growth implies economic growth (growth of GDP) which achieves significant environmental protection (Jacobs, 2012).

The tendency to conceptualise adaptation as a technical adjustment to the changing environment has gradually reduced and discourses on adaptation have been diversified (Pelling, 2011). Future scenarios for adaptation are dependent not just on modelling and quantification of human impacts, but also on societal changes (Pahl-Wostl, 2015). Social scientists, social movement activists, and practitioners started to engage in the adaptation discussion, and they put more emphasis on socio-political aspects of adaptation (Adger, 2003; Nelson et al., 2007; Tanner & Mitchell, 2008; O'Brien, 2009; Pelling, 2011).

In this context, the second type of discourse on adaptation appeared, which perceives *adaptation as a socio-political challenge* that adds to current social problems such as poverty, inequality, and unfair systems. If adaptation policy is to be successful, according to this discourse, decision-makers need to take into account that adaptation is not only about fighting against environmental changes (Adger et al., 2009). As much as hardware solutions (e.g., infrastructure building and technical support) are important, the social dimension of adaptation (fairness, equity, and participation) should be considered for 'successful' adaptation. Who are the most vulnerable people? Are they taken into account during the policymaking process? Is the implementation of the adaptation policy beneficial to them? Environmental justice comes into consideration in this regard.

On the point of 'adaptation as a socio-political challenge', green economy or green growth is not suitable for sustainable adaptation because the concepts do not challenge the inequality and injustice of the current capitalistic society (Lyon & Maxwell, 2011; Scoones et al., 2015). For example, renewable energy production and supply can worsen injustice and inequities if it focuses only on carbon emissions reduction without considering social aspects (Newell, 2015). Instead, sustainable development is often emphasised by the supporters of the second approach (Smit & Pilifosova, 2001; Swart et al., 2003).

The third discourse regards *adaptation as an opportunity to change attitudes towards the environment and ecosystems*. It criticises past industrialisation methods, which disconnected society from the ecosystem and consequently resulted in ecological degradation. The concepts related to this approach are nature-based solutions for adaptation, ecosystem-based adaptation, green infrastructure, and ecosystem-based disaster risk reduction (Kabisch et al., 2016). These concepts are based on the assumption that the conservation and restoration of ecosystem and biodiversity are critical to human well-being. Ecosystem service means "the direct and indirect contributions of ecosystems to human wellbeing" such as provisioning, regulating, habitat, and cultural services (The Economics of Ecosystems & Biodiversity, 2013). The utilisation of ecosystem services is regarded as more beneficial for social equity as well as potentially being cost-efficient.

In this discourse, the focus is still on the adaptation of human society rather than nature conservation. The main argument is that by reducing ecosystem vulnerability, the resilience of people can be increased (IUCN, 2017). Actors who hold this discourse try to attract "the commitment of governments" (Brink et al., 2016, p. 112; Chong, 2014). Together with nature conservation workers and environmental groups, many natural and social scientists also recognise the effectiveness of ecosystem services. Policy actions based on this discourse may include green roofs and walls, parks and street trees in cities, sustainable agriculture such as crop diversification, and integrated water management (Kabisch et al., 2016; IUCN, 2017).

Although ecosystem services have been used throughout human society, this discourse appeared in the discussion regarding adaptation later than the other two approaches. For example, the concept of ecosystem-based adaptation emerged as a concept around 2008 (Mercer et al., 2012; Brink et al., 2016). First, it was discussed in the context of developing countries where people have more direct contact with the environment and local knowledge of their surroundings. The discourse has been expanded to developed countries and green-blue infrastructure, along with other ecosystem service concepts began to appear in policies such as urban planning (Brink et al., 2016; Kabisch et al., 2016; Goddard et al., 2010).

Sustainable adaptation proposed by Eriksen et al. (2011) is similar to the third discourse. They differentiate sustainable adaptation from "a reformist view of sustainable development" by describing it as a concept that considers the broader effects of adaptation acts on other societies and ecosystems. Actors who can drive sustainable adaptation are not necessarily experts or workers in high positions within a government. This is a clear difference from the two other discourses.

Table 3.1 summarises the three discourse categories. The proponents of different discourses can conflict with each other during a policymaking

Table 3.1 Three discourses on climate change adaptation based on O'Brien et al. (2007)

	Adaptation as technical-fix	Adaptation as socio-political challenge	Adaptation as ecosystem service recognition
Emergence	Most common since the early period of adaptation	After the early bias towards technology eroded	Later than the other two discourses
Means of adaptation	Engineering, technical advancement	Power re-structuring, political-economy regime reform	Ecosystem service utilisation, nature restoration
Main actors/ proponents	Natural scientists, engineers, politicians, business and finance sector actors	Social scientists, practitioners, social activists	Environmental groups, conservation activists, natural/ social scientists
Narratives and rhetoric	Technology optimism, green growth	Environmental justice, sustainable development	Ecosystem service, green-blue infrastructure

process. Sometimes, a specific discourse is dominant, while policymakers disregard other discourses. This book investigates how and why different discourses regarding adaptation appear in the water management sector (Sect. 3.2.3). The approaches to solving water management issues can be categorised in a similar way to adaptation discourses. The similarity between water management discourses and adaptation discourses shows the necessity for considering adaptation policy within the historical context.

3.1.5 Resistance, Transition, and Transformation

Because adaptation refers to not only the result of an action but also the process behind it, it is possible to classify adaptation actions another way: by the degree of change. This categorisation is different from that of the previous section because it is based on the scope and range of change. Particularly, Pelling's (2011) typology of adaptation focuses on the 'process' of change, which is at the core of the definition of adaptation: resistance,[1] transition, and transformation. He asserts that "each form can include changes to values, institutions, behaviour and assets" (p. 50). This book applies Pelling's typology to the cross-case analysis and discussion chapter (Chap. 6) when the flood risk management policies of Germany and South Korea are compared.

The first type of adaptation, adaptation as '*resistance*', refers to measures to maintain existing practices and pursue changes from within the current power structure of society (Pelling, 2011). This type of adaptation is the narrowest, most minimal, and the slowest in terms of change. Pelling (2011) sees this kind of adaptation in societies where the elite has strong power to manipulate economies or the political system. This approach is attractive as a solution to vested interests in society because it does not intend to challenge the existing institutions and dominant

---

[1] In *Adaptation to Climate Change: From Resilience to Transformation* (2011), Pelling named this type of adaptation as 'resilience'; however, in another article written by Pelling, O'Brien, and Matyas (2015), this approach was named as 'resistance', not 'resilience'. In a personal communication in June 2018 in Cape Town, he explained that he changed the word because he intended those two articles for different readers and also his ideas had developed since the first book. This study uses the term 'resistance', following his later publication.

values. Actors pursuing this type of adaptation often try to maintain the status quo by denying the need for change. Resistant systems can be stable and possibly avoid risks related to change, but the inflexibility can jeopardise the system's sustainability when new circumstances emerge. Technical and organisational innovations usually fall into this type of adaptation. They are quicker to implement than deeper changes in power systems, and politically less challenging (Pelling, 2011).

The result of the policies based on resistance can be unsustainable in the long term because the policies can allow socially unjust or environmentally harmful practices. Interventions for adaptation may reinforce power structures, as Nelson and Finan (2009) explained with the drought policy of Brazil. They concluded that the government patronage during droughts without investment in capacity-building programmes created more vulnerability among the poor population (Nelson & Finan, 2009). Furthermore, if the policy focus is narrowed just to reducing climate risk, other environmental problems resulting from specific technologies or infrastructural changes may be neglected (Eriksen et al., 2011, 2015; Naess et al., 2005).

The second approach, the '*transition*' type of adaptation, focuses on the governance regime and changes within the regime (Pelling, 2011). This approach can be seen as an intermediary level of engagement between resistance and transformation. Transitional adaptation can be innovative but pursued within the prevailing order. 'Change at the margins' in Handmer and Dovers' (1996) classification of resilience is comparable to this approach. The actors acknowledge the need for change, but the extent of changes undertaken is limited to treating symptoms, not solving the root causes of risk. In addition, changes are often incremental. Adaptation activities do not threaten governing authorities, and this is probably the reason why many leaders, politicians, and business people favour this approach. Pelling (2011) points out that 'change at the margins' is the most common response to environmental problems. Minor changes may delay treating major changes essential for long-term sustainability, while a 'false sense of security' spreads out in the system (Handmer & Dovers, 1996).

In 'transitional' adaptation, social dimensions are important to advancing technical solutions for adaptation. This approach focuses on solutions within the dominant system, rather than challenging the political power structure. Citizens' participation in the decision-making process is valued and encouraged, but it does not necessarily mean that radical changes are sought. Although power structures are taken into account in the decision-making process, transitional adaptation measures are usually pursued within the prevailing order.

The third type of adaptation is *'transformation'* and represents the most profound and most radical form of adaptation. This approach requires reform in over-arching political-economic systems in order to adapt to a fundamentally new circumstance. While transitional adaptation involves actions to maintain the integrity of existing systems, transformational adaptation represents actions that go beyond the limits of transitional adaptation (Pelling et al., 2015). 'Openness and adaptability' type of resilience features tackling underlying causes of risk, showing maximum flexibility, and being prepared for unexpected threats (Handmer & Dovers, 1996). To lead transformation requires novel and innovative ways to break down the old paradigm's resistance and encourage social change (Folke et al., 2010). While the actions can substantially change the power structure in order to create long-term sustainability, there is a danger that the change may take the wrong track (mal-adaptation) and be economically inefficient.

Transformational change is infrequent in reality because of large-scale inertia in society, although it does occur from time to time (Pelling, 2011). In fact, environmental crises have not been the obvious driver for large-scale transformation in history (Newell, 2015). Handmer and Dovers (1996) argue that vested interests are the major constraint of radical changes as they are disadvantaged by the changes. It is a paradox that the state which is supposed to initiate changes for the common good is often bound together by vested interests. Some researchers point out that many climate change adaptation policies offer responses to the outcomes of vulnerability while the social and political power relations are unchallenged (Eriksen et al., 2011; Nagoda, 2015).

Researchers and policymakers have increasingly recognised that climate vulnerability cannot be solved without addressing the fundamental cause of vulnerability, which comes from the current social system (Eriksen et al., 2015). In the environmental literature, attention to transformation is growing due to another reason: the rapid speed of environmental change. As Scoones et al. (2015) point out, "there is a sense of urgency" in the debates regarding environmental changes and "radical shifts to sustainable practices" is requested (pp. 2–3). Reflecting the growing interest in transformation, the most recent IPCC Assessment Report (AR5) on adaptation dealt with a discussion on transformation. The IPCC (2014a) defines transformation as "a change in the fundamental attributes of natural and human systems" (p. 40).

Table 3.2 summarises the three types of adaptation actions by the degree of change. In a policymaking process, the proponents of different adaptation types are often in conflict with each other. Policy choices for one type of adaptation can block opportunities for another type of adaptation (Pelling et al., 2015). For example, Marshall et al. (2012) found that the attachment to place and work among peanut farmers in Queensland, Australia could lead to incremental adaptation but act as barriers for transformative adaptation. Also, Eriksen et al. (2011) emphasise that "strategies or policies that make sense from one perspective, or for one group, may at the same time reduce the livelihood viability or resource access of other groups" (p. 8). This means that a particular policy can be regarded as a 'resistance' by actors with certain positions, while others view the same policy as a 'transition' or 'transformative' adaptation depending on the standpoint.

3.2 Climate Change and Water Systems

3.2.1 Climate Change Impact on the Water Sector

Water management has always been about managing variability. As climate change adds considerable uncertainty to assessments of the quantity or quality of water, variability management is becoming more pronounced. The mean values of leading indicators such as water level are changing and water-related extremes are intensifying. As a result, more irregularities in the hydrological cycle are projected (Connor & Rast, 2009).

Table 3.2 Adaptation types by the degree of change (based on Pelling, 2011)

	Resistance	Transition	Transformation
Degree of change	Narrowest, minimal, and slowest change	Incremental change	Deepest and radical change
Power relations	Maintain the status-quo, politically less challenging, elitism	Seek changes within the regime	Power restructuring beyond the existing political-economic systems
Advantage	Stable, quick to implement	Popular among policy actors	Maximum level of flexibility, prepared for unexpected threats
Disadvantage	Inflexible, possibility of unsustainability in the long term	Minor changes may delay treating major changes	Possibility of mal-adaptation, inertia in society

Despite the high level of uncertainty about future climate scenarios, scientists generally agree that one trend in water management is inevitable due to climate change: "stationarity is dead" (Milly et al., 2008, p. 573). This statement indicates that past data on the mean states and variability cannot serve as guidance for planning and assessment in the future (Wilby & Dessai, 2010; OECD, 2013; Kundzewicz et al., 2007). Furthermore, the spatial and temporal distribution of precipitation is projected to shift, resulting in some regions becoming wetter and others drier (OECD, 2013).

The intensification of the water cycle due to warmer climate may result in more frequent and intense extreme weather events such as tropical storms, hurricanes, and typhoons (Huntington, 2006; OECD, 2013). Consequently, *flood* magnitude and frequency are increasing, while the level of low flows is decreasing causing *droughts* in many areas. Such changes will lead to impacts on navigation, power generation, irrigation, public water supply, pollution control, and water quality management (Arnell & Liu, 2001; Connor & Rast, 2009). *Freshwater* provision may also be interrupted.

3.2.1.1 Flood

The term 'floods' represents a certain image such as a submerged village, but it is a comprehensive concept which refers to various phenomena. The Special Report on Managing the Risks of Extreme Events and Disasters to Advance Climate Change Adaptation (SREX) published by the IPCC (2012) defines floods as "the overflowing of the normal confines of a stream or other body of water or the accumulation of water over areas that are not normally submerged" (p. 175). There are various types of floods including river, flash, urban, pluvial, sewer, coastal, and glacial lake outburst floods. Floods depend on various conditions such as precipitation intensity, water volume, timing, and antecedent conditions of rivers and their basins (Kundzewicz et al., 2007).

Climate change projections consistently show that warmer climates are very likely to lead to more extreme precipitation (IPCC, 2021). Milly et al. (2002) state that in most of the large river basins worldwide,

100-year peak volumes are projected to occur more frequently. They may now occur even every 2–5 years in some areas. Mid- and high-latitude regions will be affected more than other regions due to increasing mean precipitation (IPCC, 2007).

The impacts of floods on societies range from individual well-being to macro-economy. Floods can damage built environments significantly and affect many victims in a relatively short time, once they begin. In Africa, Asia, and Europe, floods are one of the most frequently reported natural disaster events. It is estimated that every year about 140 million people are affected across the globe (WDR, 2003). Flooding is linked to communicable diseases transmitted by water as well as mental health (European Environment Agency, 2016). According to the European Environment Agency, (2016), coastal flooding due to sea-level rise could potentially increase mild depression levels in the EU.

The economic damages caused by floods are tied not just to the intensity of weather events but also to socio-economic factors. The increasing number of vulnerable populations, growing wealth, and land-use changes has greatly influenced flood damages (IPCC, 2007). Particularly, floodplain areas that are prone to floods have been developed for residential, industrial, and agricultural uses around the world. According to Kron and Berz (2007), the number of large-scale floods doubled between 1950 and 2005, while economic losses rose by five times over that period.

3.2.1.2 Drought and Freshwater Provision

There are several types of droughts: (1) when precipitation is below average, meteorological drought can occur; (2) when water levels and flows are low in rivers, lakes, and groundwater, the risk of hydrological drought increases; (3) when soil moisture is low, agricultural drought can develop; and finally, (4) environmental drought refers to a combination of the above (IPCC, 2007). Although low precipitation is understood to be the leading cause of droughts, human activities such as land-use change, water demand increase, over-farming, and excessive irrigation are also important factors.

Nevertheless, scientists generally agree that climate change increases the possibility of drought. The IPCC (2014a) indicates that droughts are likely to occur more frequently in presently dry regions on low and mid-latitude continents. The frequency, duration, and severity of droughts (meteorological and hydrological) are projected to increase in most of Europe until the end of the twenty-first century (EEA, 2017).

The impacts of drought on social and ecological systems can be severe. More frequent, prolonged, and severe droughts will lead to freshwater provision issues. Usually, droughts occur in conjunction with high temperatures. The combination of high temperature and low oxygen levels in water bodies can result in the death of fish and macroinvertebrates (Bond et al., 2008). The risk of toxic algal blooms increases and this can lead to further fish deaths. Low flow of rivers can also cause habitat loss in aquatic ecosystems (Bond et al., 2008). The importance of integrating flood risk management with ecosystem conservation within river management is therefore increasing.

Ecosystem deterioration and lack of freshwater due to droughts have impacts on the economy as well as people's health and safety. Farmers are directly affected by drought because of related crop failure. Consequently, the price of agricultural products can increase, impacting the economy, as pointed out by the National Oceanic and Atmospheric Administration (NOAA) (n.d.). Low water conditions also affect the shipping and transport industries that use canals. Drought may further affect human health and safety in terms of anxiety and depression related to economic losses, and increased conflicts over water (NOAA, n.d.).

3.2.1.3 Climate Change as an Additional Stressor

Climate change is an additional strain adding to already pressing stressors to water management (Connor & Rast, 2009). Human pressures for water extraction are growing while supplies are decreasing (Palmer et al., 2008). Stressors include land-use change, urbanisation, population growth, and chemical pollution.

Social conditions heavily influence floods and droughts. Intense and long-lasting precipitation can cause dam breaks. Similarly, human encroachment into flood plains without flood response plans increases damage potential (IPCC, 2007). If policies are well designed and implemented, the degree of extreme weather impact can be significantly reduced. In developing countries and regions without enough capacity, climate change will be an additional challenge. Existing problems such as water stress, migration, food production, and international trade can be affected by extreme weather, and these phenomena can worsen in a short time frame (Bullock et al., 2009).

3.2.2 Water Management Policy Reactions to Adaptation

As the hydrological cycle is directly influenced by climate change, water management has become an essential part of adaptation plans and strategies. The water management sector needs to take climate change into account in policymaking (Connor & Rast, 2009). Awareness of the importance of climate change adaptation in the water sector rose sharply in the late 2000s (Pahl-Wostl, 2015). In many countries, climate change adaptation is now part of water resource management policy and planning and various adaptation methods have been practised from both water demand and supply perspectives (Krysanova et al., 2010). Thus, it is useful to understand how water management policies have changed since adaptation became a policy agenda (OECD, 2013; Krysanova et al., 2010).

Water management has always involved adapting to changing conditions such as increasing demand, climate variability, industrialisation, and land-use change (Krysanova et al., 2010; Arnell & Liu, 2001). Water management technologies and regulations have changed over time, and water managers have had to adapt. Krysanova et al. (2010) emphasise that the crucial difference between climate change adaptation and previous challenges is the fast speed of the change.

The end of the stability and stationarity principle in water management due to climate change presents an unprecedented challenge (Milly et al., 2008; Connor & Rast, 2009; Brown, 2010). As water cycles have become more unpredictable, historical information on water is losing its relevance. The frequencies of floods can no longer be predicted based on historical data. Policymakers and practitioners need to make decisions under increasing uncertainty (Connor & Rast, 2009; Krysanova et al., 2010). Another significant challenge is that a method which is successful in the short term could be inappropriate for long-term adaptation (Moser & Ekstrom, 2010). Meanwhile, certain policies require an extensive time scale which can be considered ineffective in the short term. For example, existing policies in urban planning, building codes, and water infrastructure planning cannot be easily adjusted in a couple of years in order to follow climate change predictions (Hallegatte, 2009). As a result, dynamic, flexible, and adaptive approaches are gaining importance instead of the previous static design paradigm (OECD, 2013).

In the field of flood prevention, there are many existing policy options: land use planning, natural water retention, technical flood protection, building regulations, insurance, warning systems, behavioural change through education, and defence systems (LAWA, 2008). These options are often divided into structural and non-structural measures. While structural measures refer to technical or engineering methods such as dams and dikes, non-structural measures include social infrastructure and population policies such as flood forecasting, warning systems, floodplain zoning, and flood-proofing. Flood risk management methods can also be divided into grey infrastructure and green-blue infrastructure. Grey infrastructure implies artificial hard infrastructures such as dams, embankments, and dikes, whereas the concept of green and blue infrastructure refers to the natural, semi-natural, and artificial networks of ecosystems such as waterways, green space, and trees (Perini, 2017).

Given the uncertain future of environmental change, the capability to change (adaptability) is vital for the water management system (Huitema & Meijerink, 2009). For enhancing adaptability to climate change, 'soft solutions' such as non-structural and green and blue infrastructure have emerged as useful methods for flood management in industrialised countries since the 1990s (Kaczmarek et al., 1996). For example, Austria promotes water retention in the catchment and the reactivation of natural flood plains in view of precautionary land use, according to its Climate Change Adaptation Strategy published in 2012 (Federal Ministry of Agriculture, Forestry, Environment and Water Management, 2012). In

Belgium, the "blue network" has been implemented to restore Brussels' rivers and water bodies (National Climate Commission, 2010). In the United Kingdom, the GraBS (Green and Blue Space Adaptation for Urban Areas and Eco Towns) project was implemented by the Town and Country Planning Association, with a particular focus on green and blue space in urban areas (Climate-ADAPT, 2016b). The Havel polder in Brandenburg, Germany, is also designed as a retention area which can lower flood peaks in the case of extreme floods (Förster et al., 2005). Those soft solutions are considered more adaptable to climate change compared to large-scale structural infrastructures, because they have less 'lock-in' effects, which means inflexibility to changing environments (Huitema & Meijerink, 2010).

It is important to note that structural measures such as dikes and dams can still play a critical role in protection from floods when they are regularly maintained and strengthened (Thieken et al., 2016). While a single dike was the most common design in the past, new approaches for re-enforcing existing dikes have been developed. For example, double dike systems and the three-step approach (spacing, broadening, and raising dikes) are considered alternative methods (Climate-ADAPT, 2016a). More intense weather events under future scenarios, however, may put technical measures, particularly dams and dikes, at higher risk of failure (Palmer et al., 2008). Thieken et al. (2016) emphasise "balanced and coordinated strategies" combining structural and non-structural approaches for flood control, pointing out that actual enforcement is still strongly biased towards structural flood defences.

3.2.3 Programmatic Discourses on Water Management

This section focuses on programmatic discourses in the water sector with the primary emphasis on flood control measures. In the water management literature, four main discourses could be identified: engineering-development, participation-equity, conservation-ecosystem service, and integrated water resource management (IWRM).

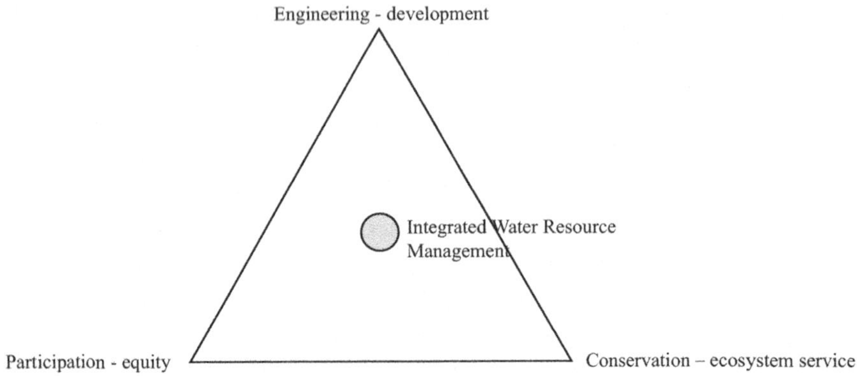

Fig. 3.2 Discursive space of water management

The first three discourses can be thought of in relation to a triangular space of water management, because they represent the main approaches of the sector (Fig. 3.2). Within the space, various policy discourses regarding water management are located. Particularly, the last discourse, the IWRM can be located in the middle of the triangle meaning that this discourse promotes the balance of economic interest, social welfare, and ecosystem protection (Global Water Partnership, 2011).

The *Engineering-development discourse* has been the most common programmatic discourse in water management. The underlying assumption of this approach is that water systems can be controlled and predicted (Trottier, 2004). Indeed, technological advancement resulted in cost-effective and efficient management in many parts of the water sector (Connor & Rast, 2009). Today's high level of water resource management such as flood protection, hydropower generation, and waterway construction could be achieved through the engineering approach (Huitema & Meijerink, 2010; Huntjens et al., 2011; Pahl-Wostl, 2015). Because floodplain development, navigation, and hydropower have contributed to economic growth, engineering-based water management and development are often linked together.

Most engineering and technological innovations have both positive and negative effects on the environment, even when those effects may not be intended (Connor & Rast, 2009). For example, dike heightening and reinforcement can prevent floods but the ecosystem around the project site's landscape can be negatively affected. In many countries including Germany and South Korea, environmental impact assessments are compulsory before major infrastructure construction. Forced resettlement or the alteration of the ecosystem caused by dams also exemplify a side effect of the engineering approach.

Structural infrastructures such as dikes and dams have strong supporters and opponents at the same time (Climate-ADAPT, 2016a). Stakeholders of engineering-based projects such as civil engineers, hydrologists, construction businesses, and developers tend to support large-scale engineering projects because of their perceived economic benefits. Maass (1951) described how engineers, politicians, and other major interest groups closely collaborated on local river development projects in the United States in the 1940s. The traditional engineering approach, however, has been increasingly criticised in industrialized countries since the 1970s as concerns for the environment rose. Nowadays, large dams are opposed not only by environmentalists but also by citizens in many countries because of their adverse effects on ecology and landscape. Given the challenges of climate change, technological fixes that work now may be considered inappropriate in the future (Pahl-Wostl, 2015).

The engineering-development discourse is still globally popular, and many water management projects follow this approach. The number of dams around the world is increasing due to the boom of hydropower. In Eastern Europe, conflicts have arisen regarding dam construction because of the 2800 dams planned across the Balkans (Neslen, 2018). One study revealed that two-thirds of great rivers in the world do not flow freely anymore due to the construction of structural infrastructures along the rivers (Grill, 2015).

The second programmatic discourse emphasises *nature conservation and ecosystem services*. As indicated in Sect. 3.1.4, ecosystem services are the "conditions and processes of ecosystems" which sustain and enhance human life (Daily, 1997, p. 3). The basic assumption of this discourse is that restoring or keeping the natural state of water bodies provides significant benefits to the ecosystem as well as human societies. Many environmental activists and ecologists hold this storyline. This approach has gradually gained importance since the 1970s when environmental movements earned broad support in Europe and the United States. Environmentalists argue that lessons learned from experience with grey infrastructure such as habitat and biodiversity loss should be reflected in water management (Perini, 2017; Depietri & McPhearson, 2017).

The advocates of this discourse assert that technical solutions have a disadvantage from a long-term perspective. For example, high dikes increase the maximum water level and subsequently, the risk of more severe floods. A dam collapse results in disastrous floods in an extensive area. The side effects of traditional engineering approaches include massive-scale alterations to social-ecological systems (Huitema & Meijerink, 2009). In contrast, riparian areas and floodplains that allow rivers to flow more freely can reduce the flood peak while nature can be preserved in the floodplains (Climate-ADAPT, 2016a). The OECD (2013) states that ecosystem-based adaptation and green and blue infrastructure methods can be cost-effective, particularly for addressing uncertainty linked to climate change because lock-in effects with capital-intensive infrastructures can be avoided or delayed.

As this discourse has gained support, many dam removal projects have been completed around the world, particularly in the United States. The Wild and Scenic Rivers Act shows a clear change in the water policy moving away from grey infrastructures (US Fish and Wildlife Service, 2019). Bruce Brown's *Mountain in the Clouds: A Search for the Wild Salmon* (1982) is known as an influential book which contributed to social reflection on river ecosystems in the country. The policy change can be observed in the case of the Elwha River, where two of once the biggest dams in the United States were located; their removal was completed in 2014 in order to restore the river ecosystem (Lohan, 2018).

In Europe, one of the most notable changes in the water management sector occurred in the Netherlands. The Dutch government traditionally prioritised dikes and water pumping against flood risks, but such measures became questionable after the near-floods of 1993 and 1995 (Bauer & Steurer, 2015). After 1995, the Room for the River Programme was implemented by the Dutch government aiming for the restoration of natural floodplains in order to reduce flood risk (Rijkswaterstaat, 2016). Interviewees in the Netherlands pointed out that the background of the programme has raised awareness about the environment (M. Kaufmann, personal communication, 25 October 2018; D. Huitema, personal communication, 7 November 2018). One interviewee explained how the change in water management discourse affected Dutch hydrology engineers' activities.

> There is a long history in the Netherlands that hydrology engineers are involved in other countries to give advice. There are many areas in the 1970s for which Dutch engineers advised creating polders (with dikes and dams) everywhere. Now Dutch engineers go back to the same countries and advise their old colleagues again because they see the negative effects of the previous technology. Often, they are not sustainable. Above all, the values of society have changed. The same engineers came to different conclusions as society changed. (E. Mosselman from Deltares, personal communication, 8 November 2018)

The third programmatic discourse on water management is participation-equity. Together with the nature conservation and ecosystem service discourse, this discourse has gained more support since the 1970s as environmental movements grew in strength and number. Since then, the perception of what governments' roles in water management should be has changed from that of central actor to one of many actors in a far more complex environment (Ingram, 2011; Pahl-Wostl, 2015).

Such changes can be mainly observed in European water policies. Pahl-Wostl (2015, p. 11) states that the evolution of European water policies shows the general shift from "command and control" to more market-based and more participatory approaches. Traditionally, technology-oriented water management methods have been often planned and implemented in a centralised way through "rigid and detailed" plans (Pahl-Wostl, 2015, p. 3). Such planning processes changed as programmatic discourse became diversified and nature-based approaches emerged. The EU Water Framework Directive (WFD) (2000/60/EC, EC, 2000) is an exemplary document of such change. WFD Article 14 states that there must be "the active involvement of all interested parties in the implementation of the directive".

Because of climate change, participatory governance is gaining more importance. A command-and-control attitude can prevent adaptive management which is becoming more necessary due to the uncertainties of climate scenarios (Krysanova et al., 2010). Pahl-Wostl (2015) argues that because of the changing reality of water management, the paradigm shift from management to governance is required. While the concept 'management' implies that a few people can control water, the concept 'governance' recognises the multiplicity of actors and centres, meaning broad participation (Trottier, 2004, p. 2).

Equity is another essential value for this discourse. Although freshwater availability is the universal goal of water policy, the right to access and use freshwater is a highly political issue. Equity has become more critical because freshwater availability is decreasing in many areas of the world due to socio-environmental changes such as population growth and climate change (Basu & Shaw, 2013). Equity addresses questions such as the higher exposure of vulnerable peoples to flood risks (Johnson et al., 2007). Certain river basins or coastal areas are exposed to flood risks more frequently due to spatial planning or structural infrastructure. When the river course upper stream is straightened, flood risks for downstream areas can increase. The participation-equity discourse points out that in order to make water management policy fair and just, relevant stakeholders should be engaged in decision-making processes.

The fourth discourse, the IWRM, is a balanced stance. This concept emerged as an alternative institutional arrangement to the fragmented management practices which have dominated the water sector across the globe. The idea is that water governance, management, and use cannot be considered separately (De Gooijer et al., 2009, p. 245).

Although IWRM is interpreted in various ways, Arnell and Liu (2001) point out three main components of IWRM: (1) consideration of both supply-side and demand-side actions, (2) inclusion of all stakeholders in the decision-making process, and (3) monitoring and review of water management. This explanation is in line with the definition suggested by the Global Water Partnership (2011), which states that IWRM is "a process which promotes the coordinated development and management of water, land and related resources, in order to maximise the resultant economic and social welfare equitably without compromising the sustainability of vital ecosystems". In the programmatic discourse triangle suggested in Fig. 3.2, the IWRM is placed in the middle, where the three water management discourses are integrated and compromised.

The IWRM concept was extensively discussed in the United Nations Conference on Environment and Development held in Rio de Janeiro in 1992. Ten years later, the countries that participated in the World Summit on Sustainable Development agreed to develop national IWRM plans by 2005 (Bullock et al., 2009, p. 4). Subsequently, the 16th Session of the Commission on Sustainable Development (2008) endorsed IWRM as an essential framework for effective water resource management (De Gooijer et al., 2009, p. 241). As the IWRM concept has been promoted and spread by international organisations including the EU, many countries now apply IWRM at the basin level (Bullock et al., 2009, p. 4).

The implementation of IWRM has been more challenging than expected. Based on the analysis of 67 EU projects related to IWRM, De Gooijer et al. (2009) conclude that although IWRM provides a useful framework, more guidance is needed for policy planning and implementation. Krysanova et al. (2010) criticise that IWRM is not a sufficient measure to tackle uncertainty in water management in relation to climate change.

Table 3.3 summarises the four discourses presented in this section following the discourse analysis framework suggested in Sect. 2.2.2. Although these discourses do not represent all discussions within the water management field, they are the frames commonly found in the field, particularly in the flood risk management area. The policymaking process for water management can be described as a continuous fight between those discourses. During the interviews conducted in South Korea and Germany, it was observed that interviewees frequently used more than one of those four discourses. Therefore, the cases of this book are analysed and compared with each other based on the four discourses.

The engineering-development, participation-equity, and conservation-ecosystem service discourses within water management show similarities with the discourses about climate change adaptation introduced in Sect. 3.1.4: technical-fix, socio-political task, and ecosystem service. The similarities of the water management and adaptation programmatic discourse categories can be attributed to the fact that both subjects (adaptation and water management) are influenced by general approaches to environmental problems (Fünfgeld & McEvoy, 2014). Another reason may be the overlap of the actors involved in adaptation and water management

Table 3.3 Four types of programmatic discourses on water management

		Engineering-development	Conservation-ecosystem service	Participation-equity	Integrated management
Internal elements	Frame	Command & control, practicality, feasibility	Ecosystem restoration	Social equity and civil engagement	Sustainable development
	Storyline	Technological advancement can fix problems	Ecosystem sustains human life	Fair intervention reduces risks	Economy, social welfare, and ecosystem in harmony
	Rhetoric and metaphor	Economic growth, modernity	Room for the rivers, salmon	The multiplicity of centres, fair and just society	Balanced management
External elements	Actors and discourse-coalitions	Civil engineers, hydrologists, construction business, and developers	Environmentalists and conservationists	Env. NGOs, development practitioners, and social scientists	International organisations including the UN and the EU
	Links to institutions	Centralised decision-making	Participatory, multilevel governance	Participatory governance	Coordinative institutions

policies as found during the interviews. The similarities of the two programmatic discourse categorisations provide a firm ground for the integrative analysis of water management and adaptation policies, which is pursued in this book.

3.2.4 Water Management Institutions

While discourse analysis allows overviewing diverse approaches to water management, institutional analysis helps to understand how the different discourses have emerged and influenced water policy. Arnell and Liu (2001) point out that different institutional settings have very much influenced the techniques used in the water sector in various countries. Unlike climate change adaptation, water management institutions have a long history in most countries. This section focuses on flood and drought management institutions and looks at how institutional models have been changed and developed as well as how climate change is challenging water management institutions.

3.2.4.1 Institutional Changes in Water Management since 1950

The water sector has experienced various institutional settings: hierarchical centralisation, coordinated river basin planning and management, devolution and decentralisation, and markets and privatisation (Ingram, 2011). In modern society, the prediction and control approach has been a widely practised formal institution. It is often regarded as the traditional water management practice (Huntjens et al., 2011; Pahl-Wostl, 2015). 'Prediction and control' became essential during the nineteenth century as urban areas with dense population increased and land use intensified (Pahl-Wostl, 2015). This institutional model was usually practised in a top-down manner by centralised governments (Pahl-Wostl, 2015).

Changes in institutional settings started to appear in the late 1980s. According to Pahl-Wostl (2015), the market became a key instrument for water management because of the decentralisation and privatisation of the sector. The World Bank was active in leading this change (World

Bank, 1993). Authority and responsibility for water management were transferred from national governments to sub-national governments and the private sector (Pahl-Wostl, 2015). While this trend could be mainly found in the water supply and sanitation sector, it has also affected overall water management policy (Bakker, 2010).

Another significant change in water management institutions was the rise of integrative approaches. Until the 1990s (and in some countries still today), subsector specialists worked independently and hardly interacted with each other (Bullock et al., 2009). Over time, however, the fragmentation of the sector became increasingly problematic as the pressure on water resources grew with industrialisation and land-use change (UNESCO, 2009). Based on the experience of practitioners, IWRM emerged as an alternative.

Actors involved in the policy process of water management have been diversified as a result of the urbanisation, democratisation, and privatisation of the sector. Actors include governments at various levels, other sector agencies, river basin commissions, water resource authorities, consumer organisations, private companies, and civil organisations (Björklund et al., 2009). The role of actors becomes "blurred in more complex and intertwined governance regimes" compared to traditional water governance (Pahl-Wostl, 2009, p. 357).

In the late twentieth century, discussions at the international and regional levels started to provide directions, norms, and tools. The Stockholm Declaration of 1972 is referred to as the beginning of global policy frameworks for water (Björklund et al., 2009). The United Nations Economic Commission for Europe Convention on the Protection and Use of Transboundary Watercourses and International Lakes serves for water protection in the participating countries (Björklund et al., 2009). Most recently, the UN Convention on the Law of the Non-navigational Uses of International Watercourses entered into force in 2014 more than 17 years after its adoption by the UN General Assembly. Since the EU WFD was adopted, transboundary waters in the EU are managed by international or bilateral commissions for river basins or lakes (UBA, 2013). In sum, water management institutions have become more complex with multiple levels of participation from local to international governance (Kaika, 2004).

Despite these institutional changes in the water management sector since the twentieth century, national governments continue to play a critical role (Huitema & Meijerink, 2010). National governments can control lower levels of government, sanction private governance and influence international organisations (Pierre & Peters, 2005; Huitema & Meijerink, 2010). In the two cases investigated in this study, national-level policy has been influential although the extent of the national governments' influence was different.

3.2.4.2 Climate Change as a Challenge to Water Management Institutions

Because of the high level of uncertainty and the unprecedented rate of climate change, water managers cannot simply rely on historical experience and data anymore (OECD, 2013). There is, however, a 'lock-in', associated with water policy which stems in part from the dominance of established technologies (Arthur, 1994). In institutional analyses, this is referred to as 'path dependence'. For example, once a dam is built, it is not easy to demolish the dam within a short period because of the effects the sudden change would bring to society and the ecosystem. Pahl-Wostl (2009) indicates that resource governance systems are often the result of long-time evolution and they are intertwined with infrastructures. Such interdependence of elements in resource governance systems can hinder innovative changes (Huntjens et al., 2011).

Nevertheless, disastrous events that result in serious socio-economic damage have often been the main trigger for changes in the water management paradigm (Huitema & Meijerink, 2010; Krysanova et al., 2010). Floods, hurricanes, typhoons, and prolonged droughts have altered national policies. For example, the Mississippi floods in the United States in 1993 and the European floods in 1994–1995 served as a momentum for major reviews of flood policies (Kaczmarek et al., 1996). Experts generally agreed that climate-related disasters are one of the significant drivers of developing climate change adaptation strategies in river basins (Krysanova et al., 2010).

Some scholars hold the view that currently, most water management systems have limited capacity to deal with climate change (Connor & Rast, 2009; De Gooijer et al., 2009; Huntjens et al., 2011; Pahl-Wostl, 2015). De Gooijer et al. (2009) argue that some institutional settings such as political power monopolies, unilateral steering by government and bureaucracy, hierarchical control, top-down management, and institutional fragmentation can hinder adaptation to current and future challenges. Pahl-Wostl (2015) also states that 'command and control' measures in a top-down manner could be useful in a stable environment but they might not be suitable in non-linear change and unprecedented systems.

When the implications of climate change are taken into account, scholars suggest that it is necessary to transform the water management approach to adaptive water management, which is defined as a systematic improvement process by learning from previous policy implementation (Huntjens et al., 2011; Krysanova et al., 2010; Arnell & Liu, 2001; Pahl-Wostl, 2007, 2009; Folke et al., 2005). In this way, the impact of uncertain conditions due to climate change can be reduced (Krysanova et al., 2010).

In the next two chapters (Chaps. 4 and 5), the adaptation and water management policies of the two cases of this book, South Korea and Germany, are analysed. The categorisations of programmatic discourses on adaptation and water management introduced in this chapter are applied in the analysis of the case studies. The relevant institutions of each case are also explained based on the background information provided in this chapter.

References

Adaptation Fund. (2017). *Concept note for a regional project.* Retrieved February 25, 2020, from https://www.adaptation-fund.org/wp-content/uploads/2017/04/ADAPT-WAP-CN-OSS-AF-April2017.pdf

Adger, W. N. (2000). Institutional adaptation to environmental risk under the transition in Vietnam. *Annals of the Association of American Geographers, 90*(4), 738–758.

Adger, W. N. (2003). Social capital, collective action and adaptation to climate change. *Economic Geography, 79*, 387–404.

Adger, W. N., Dessai, S., Goulden, M., Hulme, M., Lorenzoni, I., Nelson, D., Naess, L. O., Wolf, J., & Wreford, A. (2009). Are there social limits to adaptation to climate change? *Climate Change, 93*, 335–354.

Andersson, E., & Keskitalo, E. C. H. (2018). Adaptation to climate change? Why business-as-usual remains the logical choice in Swedish forestry. *Global Environmental Change, 48*, 76–85. https://doi.org/10.1016/j.gloenvcha.2017.11.004

Arnell, N., & Liu, C. (2001). Hydrology and water resources. In J. McCarthy, O. Canziani, N. Leary, D. Dokken, & K. White (Eds.), *Climate change 2001, impacts, adaptation, and vulnerability* (pp. 193–227). Intergovernmental panel on climate change). Cambridge University Press.

Arthur, W. B. (1994). *Increasing returns and path dependence in the economy*. University of Michigan Press.

Ayers, J. M., & Huq, S. (2009). The value of linking mitigation and adaptation: A case study of Bangladesh. *Environmental Management, 43*(5), 753–764. https://doi.org/10.1007/s00267-008-9223-2

Bakker, K. (2010). *Privatizing water: Governance failure and the World's urban water crisis*. Cornell Univ.

Barnett, J., Evans, L. S., Gross, C., Kiem, A. S., Kingsford, R. T., Palutikof, J. P., Pickering, C. M., & Smithers, S. G. (2015). From barriers to limits to climate change adaptation: Path dependency and the speed of change. *Ecology and Society, 20*(3), 5. https://doi.org/10.5751/ES-07698-200305

Basu, M., & Shaw, R. (2013). Water policy, climate change and adaptation in South Asia. *International Journal of Environmental Studies, 70*(2), 175–191. https://doi.org/10.1080/00207233.2013.781736

Bauer, A., Feichtinger, J., & Steurer, R. (2012). The governance of climate change adaptation in 10 OECD countries: Challenges and approaches. *Journal of Environmental Policy & Planning, 14*(3), 279–304. https://doi.org/10.1080/1523908X.2012.707406

Bauer, A., & Steurer, R. (2015). National adaptation strategies, what else? Comparing adaptation mainstreaming in German and Dutch water management. *Regional Environmental Change, 15*(2), 341–352. https://doi.org/10.1007/s10113-014-0655-3

Biagini, B., Bierbaum, R., Stults, M., Dobardzic, S., & McNeeley, S. M. (2014). A typology of adaptation actions. A global look at climate adaptation actions financed through the global environment facility. *Global Environmental Change, 25*, 97–108. https://doi.org/10.1016/j.gloenvcha.2014.01.003

Biesbroek, G. R., Swart, R. J., Carter, T. R., Cowan, C., Henrichs, T., Mela, H., et al. (2010). Europe adapts to climate change: Comparing national adaptation strategies. *Global Environmental Change, 20*(3), 440–450. https://doi.org/10.1016/j.gloenvcha.2010.03.005

Bijlsma, L., Ehler, C. N., Klein, R. J. T., Kulshrestha, S. M., McLean, R. F., Mimura, N., Nicholls, R. J., Nurse, L. A., Perez Nieto, H., Stakhiv, E. Z., Turner, R. K., & Warrick, R. A. (1996). Coastal zones and small islands. In R. T. Watson, M. C. Zinyowera, & R. H. Moss (Eds.), *Climate change 1995. Impacts, adaptations and mitigation of climate change: Scientific-technical analyses. Contribution of working group II to the second assessment report of the intergovernmental panel on climate change* (pp. 289–324). Cambridge University Press.

Björklund, G., Burchi, S., Connor, R., Cosgrove, W., Hendry, S., Moriarty, P., Rast, W., Salame, L., & Winpenny, J. (2009). Policies, laws and finance. In World Water Assessment Programme (Ed.), *The United Nations world water development report 3: Water in a changing world* (pp. 49–67). UNESCO & Earthscan.

Bond, N. R., Lake, P. S., & Arthington, A. H. (2008). The impacts of drought on freshwater ecosystems: An Australian perspective. *Hydrobiologia, 600*(1), 3–16. https://doi.org/10.1007/s10750-008-9326-z

Boykoff, M. T. (2011). *Who speaks for the climate? Making sense of media reporting on climate change*. Cambridge University Press.

Boykoff, M. T., Ghosh, A., & Venkateswaran, K. (2013). Media coverage of discourse on adaptation. In S. C. Moser & M. T. Boykoff (Eds.), *Successful adaptation to climate change. Linking science and policy in a rapidly changing world* (pp. 237–252). Routledge.

Brink, E., Aalders, T., Ádám, D., Feller, R., Henselek, Y., Hoffmann, A., et al. (2016). Cascades of green: A review of ecosystem-based adaptation in urban areas. *Global Environmental Change, 36*, 111–123. https://doi.org/10.1016/j.gloenvcha.2015.11.003

Brown, B. (1982). *Mountain in the clouds*. Simon and Schuster.

Brown, C. (2010). The end of reliability. *Journal of Water Resources Planning and Management, 136*(2), 143–145. https://doi.org/10.1061/(ASCE)WR.1943-5452.65

Bullock, A., Cosgrove, W., van der Hoek, W., & Winpenny, J. (2009). Getting out of the box–linking water to decisions for sustainable development. In World Water Assessment Programme (Ed.), *The United Nations world water development report 3: Water in a changing world* (pp. 3–24). UNESCO & Earthscan.

Carter, T. R., Parry, M. L., Harasawa, H., & Nishioka, S. (1994). *IPCC technical guidelines for assessing climate change impacts and adaptations*. IPCC.

Chong, J. (2014). Ecosystem-based approaches to climate change adaptation: Progress and challenges. *International Environmental Agreements, 14*, 391–405.

Climate-ADAPT. (2016a). *Adaptation or improvement of dikes and dams* (2015).

Climate-ADAPT. (2016b). *GraBS green and blue space adaptation for urban areas and eco towns*. Retrieved February 7, 2020, from https://climate-adapt.eea.europa.eu/metadata/projects/green-and-blue-space-adaptation-for-urban-areas-and-eco-towns

Climate Change Adaptation Resource Center. (2019). *Strategies for climate change adaptation*. Retrieved February 27, 2020, from https://www.epa.gov/arc-x/strategies-climate-change-adaptation

Connor, R., & Rast, W. (2009). Technological Innovation. In World Water Assessment Programme (Ed.), *The United Nations world water development report 3: Water in a changing world* (pp. 41–48). UNESCO & Earthscan.

Crutzen, P. J. (2006). The "Anthropocene". In E. Ehlers & T. Krafft (Eds.), *Earth system science in the Anthropocene*. Springer.

Daily, G. C. (1997). *Nature's services*. Societal dependence on natural ecosystems.

De Gooijer, G., Rast, W., & Tropp, H. (2009). Options inside the water box. In World Water Assessment Programme (Ed.), *The United Nations world water development report 3: Water in a changing world* (pp. 241–268). UNESCO & Earthscan.

Depietri, Y., & McPhearson, T. (2017). Integrating the Grey, green, and blue in cities: Nature-based solutions for climate change adaptation and risk reduction. In N. Kabisch, H. Korn, J. Stadler, & A. Bonn (Eds.), *Nature-based solutions to climate change adaptation in urban areas*. Springer.

Dupuis, J., & Biesbroek, R. (2013). Comparing apples and oranges: The dependent variable problem in comparing and evaluating climate change adaptation policies. *Global Environmental Change, 23*(6), 1476–1487. https://doi.org/10.1016/j.gloenvcha.2013.07.022

Eriksen, S., Aldunce, P., Bahinipati, C., Martins, R. D., Molefe, J. I., Nhemachena, C., et al. (2011). When not every response to climate change is a good one. Identifying principles for sustainable adaptation. *Climate and Development, 3*(1), 7–20. https://doi.org/10.3763/cdev.2010.0060

Eriksen, S., Nightingale, A. J., & Eakin, H. (2015). Reframing adaptation. The political nature of climate change adaptation. *Global Environmental Change, 35*, 523–533. https://doi.org/10.1016/j.gloenvcha.2015.09.014

European Commission. (n.d.). *"Supporting climate action through the EU budget"*. Retrieved February 6, 2020, from https://ec.europa.eu/clima/policies/budget/mainstreaming_en

European Commission. (2000). *Directive 2000/60/EC of the European Parliament and of the Council of 23 October 2000 establishing a framework for community action in the field of water policy* (Water Framework Directive).
European Environment Agency. (2014). *National adaptation policy processes in European countries–2014*. Edited by European Environment Agency.
European Environment Agency. (2016). *Floods and health*. Retrieved January 5, 2020, from https://www.eea.europa.eu/data-and-maps/indicators/floods-and-health-1/assessment
European Environment Agency. (2017). *Meteorological and hydrological droughts*. Retrieved February 7, 2020, from https://www.eea.europa.eu/data-and-maps/indicators/river-flow-drought-2/assessment
Federal Ministry of Agriculture, Forestry, Environment and Water Management, Austria. (2012). *The Austrian strategy for adaptation to climate change*. Federal Ministry of Agriculture, Forestry, Environment and Water Management.
Folke, C., Hahn, T., Olsson, P., & Norberg, J. (2005). Adaptive governance of social-ecological systems. *Annual Review of Environment and Resources, 30*(1), 441–473. https://doi.org/10.1146/annurev.energy.30.050504.144511
Folke, C., Carpenter, B. H., Walker, M., Scheffer, F. S., Chapin, I. I. I., & Rockström, J. (2010). Resilience thinking: Integrating resilience, adaptability and transformability. *Ecology and Society, 15*(4).
Förster, S., Kneis, D., Gocht, M., & Bronstert, A. (2005). Flood risk reduction by the use of retention areas at the Elbe River. *International Journal of River Basin Management, 3*(1), 21–29. https://doi.org/10.1080/15715124.2005.9635242
Fünfgeld, H., & McEvoy, D. (2014). Frame divergence in climate change adaptation policy. Insights from Australian local government planning. *Environ Plann C Gov Policy, 32*(4), 603–622. https://doi.org/10.1068/c1234
Füssel, H. M. (2007). Adaptation planning for climate change: Concepts, assessment approaches, and key lessons. *Sustainability Science, 2*, 265–275.
Global Water Partnership. (2011). *What is IWRM?* Retrieved February 8, 2020, from https://www.gwp.org/en/GWP-CEE/about/why/what-is-iwrm/
Goddard, M. A., Dougill, A. J., & Benton, T. G. (2010). Scaling up from gardens: Biodiversity conservation in urban environments. *Trends in Ecology & Evolution, 25*(2), 90–98.
Grill, G. (2015). Dams reshape the world's rivers. *Nature, 517*(7536), 530. https://doi.org/10.1038/517530a
Hallegatte, S. (2009). Strategies to adapt to an uncertain climate change. *Global Environmental Change, 19*, 240–247.

Handmer, J. W., & Dovers, S. R. (1996). A typology of resilience: Rethinking institutions for sustainable development. *Industrial & Environmental Crisis Quarterly, 9*(4), 482–511. https://doi.org/10.1177/108602669600900403

Huitema, D., & Meijerink, S. (2009). *Water policy entrepreneurs. A research companion to water transitions around the globe.* Edward Elgar.

Huitema, D., & Meijerink, S. (2010). Realizing water transitions. The role of policy entrepreneurs in water policy change. *Ecology and Society, 15*(2).

Huntington, T. G. (2006). Evidence for intensification of the global water cycle: Review and synthesis. *Journal of Hydrology, 319*(1–4), 83–95. https://doi.org/10.1016/j.jhydrol.2005.07.003

Huntjens, P., Pahl-Wostl, C., Rihoux, B., Schlüter, M., Flachner, Z., Neto, S., et al. (2011). Adaptive water management and policy learning in a changing climate. A formal comparative analysis of eight water management regimes in Europe, Africa and Asia. *Environmental Policy and Governance, 21*(3), 145–163. https://doi.org/10.1002/eet.571

Hwang, Y. & Hwanbong, J. (2018, August 6). More severe heatwaves to vulnerable people. *Hangyeoreh.* Retrieved April 2, 2020, from http://www.hani.co.kr/arti/society/health/856346.html

Ingram, H. (2011). Beyond universal remedies for good water governance. In A. Garrido & H. M. Ingram (Eds.), *Water for food in a changing world* (pp. 241–261). Routledge.

IPCC. (2007). In S. Solomon, D. Qin, M. Manning, Z. Chen, M. Marquis, K. B. Averyt, M. Tignor, & H. L. Miller (Eds.), *Climate change 2007: The physical science basis. Contribution of working group I to the fourth assessment report of the intergovernmental panel on climate change.* Cambridge University Press.

IPCC. (2012). Managing the risks of extreme events and disasters to advance climate change adaptation: Special report of the intergovernmental panel on climate change [Field, C.B., V. Barros, T.F. Stocker, D. Qin, D.J. Dokken, K.L. Ebi, M.D. Mastrandrea, K.J. Mach, G.-K. Plattner, S.K. Allen, M. Tignor, and P.M. Midgley (Eds.)] Cambridge University Press.

IPCC. (2013). In T. F. Stocker, D. Qin, G.-K. Plattner, M. Tignor, S. K. Allen, J. Boschung, A. Nauels, Y. Xia, V. Bex, & P. M. Midgley (Eds.), *Climate change 2013: The physical science basis. Contribution of working group I to the fifth assessment report of the intergovernmental panel on climate change.* Cambridge University Press.

IPCC. (2014a). *Climate change 2014: Impacts, adaptation and vulnerability, volume 2, regional aspects. Working group II contribution to the IPCC fifth assessment report.* Cambridge University Press.

IPCC. (2014b). *Climate change 2014. Mitigation of climate change working group III contribution to the fifth assessment report of the intergovernmental panel on climate change*. Cambridge University Press.

IPCC. (2021). Summary for policymakers. In V. Masson Delmotte, P. Zhai, A. Pirani, S. L. Connors, C. Péan, S. Berger, N. Caud, Y. Chen, L. Goldfarb, M. I. Gomis, M. Huang, K. Leitzell, E. Lonnoy, J. B. R. Matthews, T. K. Maycock, T. Waterfield, O. Yelekçi, R. Yu, & B. Zhou (Eds.), *Climate change 2021: The physical science basis. Contribution of working group I to the sixth assessment report of the intergovernmental panel on climate change*. Cambridge University Press. In Press.

IUCN. (2017). *Issues brief: Ecosystem-based adaptation*. Retrieved February 6, 2020, from https://www.iucn.org/resources/issues-briefs/ecosystem-based-adaptation

Jacobs, M. (2012). *Green growth: Economic theory and political discourse*. Centre for Climate Change Economics and Policy.

Johnson, C., Penning-Rowsell, E., & Parker, D. (2007). Natural and imposed injustices: The challenges in implementing 'fair' flood risk management policy in England. *The Geographical Journal, 173*(4), 374–390. https://doi.org/10.1111/j.1475-4959.2007.00256.x

Juhola, S., Keskitalo, E., Carina, H., & Westerhoff, L. (2011). Understanding the framings of climate change adaptation across multiple scales of governance in Europe. *Environmental Politics, 20*(4), 445–463. https://doi.org/10.1080/09644016.2011.589571

Kabisch, N., Frantzeskaki, N., Pauleit, S., Naumann, S., McKenna, D., Artmann, M., et al. (2016). Nature-based solutions to climate change mitigation and adaptation in urban areas: Perspectives on indicators, knowledge gaps, barriers, and opportunities for action. *E&S, 21*(2). https://doi.org/10.5751/ES-08373-210239

Kaczmarek, Z., Amell, N. W., & Stakhiv, E. Z. (1996). Water resources management. In R. T. Watson, M. C. Zinyowera, & R. H. Moss (Eds.), *Climate change 1995. Impacts, adaptations and mitigation of climate change: Scientific-technical analyses. Contribution of working group II to the second assessment report of the intergovernmental panel on climate change* (pp. 469–486). Cambridge University Press.

Kaika, M. (2004). Water for Europe: The creation of the European water framework directive. In J. Trottier & P. Slack (Eds.), *Managing water resources past and present* (pp. 89–116). Oxford University Press.

Kron, W., & Berz, G. (2007). Flood disasters and climate change: Trends and options–a (re-)insurer's view. In J. L. Lozán, H. Graßl, P. Hupfer, L. Menzel, & C.-D. Schönwiese (Eds.), *Global change: Enough water for all?* (pp. 268–273). Wiss. Auswertungen.

Krysanova, V., Dickens, C., Timmerman, J., Varela-Ortega, C., Schlüter, M., Roest, K., et al. (2010). Cross-comparison of climate change adaptation strategies across large river basins in Europe, Africa and Asia. *Water Resources Management, 24*(14), 4121–4160. https://doi.org/10.1007/s11269-010-9650-8

Kundzewicz, Z.W., L.J. Mata, N.W. Arnell, P. Döll, P. Kabat, B. Jiménez, K.A. Miller, T. Oki, Z. Sen and I.A. Shiklomanov (2007). Freshwater resources and their management. In M.L. Parry, O.F. Canziani, J.P. Palutikof, P. J. van der Linden and C.E. Hanson, (Eds.). Climate change 2007: Impacts, adaptation and vulnerability. Contribution of working group II to the fourth assessment report of the intergovernmental panel on climate change (pp. 173–210). Cambridge University Press.

LAWA. (2008). *Strategie zur Umsetzung der Hochwasserriskomanagement-Richtlinie in Deutschland.* LAWA.

Leach, M. (2015). What is green? Transformation imperatives and knowledge politics. In I. Scoones, M. Leach, & P. Newell (Eds.), *The politics of green transformations* (pp. 25–38). Routledge.

Leary, N. A. (1999). A framework for benefit-cost analysis of adaptation to climate change and climate variability. *Mitigation and Adaptation Strategies for Global Change, 4*(3/4), 307–318. https://doi.org/10.1023/A:1009667706027

Lindseth, G. (2005). Local level adaptation to climate change: Discursive strategies in the Norwegian context. *Journal of Environmental Policy & Planning, 7*(1), 61–84. https://doi.org/10.1080/15239080500251908

Lohan, T. (2018). The Elwha's living laboratory: Lessons from the World's largest dam-removal project. In *the Revelator*. Retrieved December 10, 2021, from https://therevelator.org/elwha-dam-removal/

Luke, T. W. (2008). The politics of true convenience or inconvenient truth: Struggles over how to sustain capitalism, democracy, and ecology in the 21st century. *Environment & Planning A, 40*(8), 1811–1824. https://doi.org/10.1068/a40158

Lyon, T. P., & Maxwell, J. W. (2011). Greenwash: Corporate environmental disclosure under threat of audit. *Journal of Economics & Management Strategy, 20*(1), 3–41. https://doi.org/10.1111/j.1530-9134.2010.00282.x

Maass, A. (1951). *Muddy waters.* Harvard University Press.

Marshall, N. A., Park, S. E., Adger, W. N., Brown, K., & Howden, S. M. (2012). Transformational capacity and the influence of place and identity. *Environmental Research Letters, 7*(3), 34022. https://doi.org/10.1088/1748-9326/7/3/034022

McCarthy, J., Canziani, O., Leary, N., Dokken, D., & White, K. (Eds.). (2001). *Climate change 2001, impacts, adaptation, and vulnerability*. Cambridge University Press.

Mercer, J., Kelman, I., Alfthan, B., & Kurvits, T. (2012). Ecosystem-based adaptation to climate change in Caribbean small island developing states: Integrating local and external knowledge. *Sustainability, 4*, 1908–1932.

Millennium Ecosystem Assessment. (2005). *Ecosystems and human well being: Synthesis*. Island Press.

Milly, P. C. D., Wetherald, R. T., Dunne, K. A., & Delworth, T. L. (2002). Increasing risk of great floods in a changing climate. *Nature, 415*(6871), 514–517. https://doi.org/10.1038/415514a

Milly, P. C. D., Betancourt, J., Falkenmark, M., Hirsch, R. M., Kundzewicz, Z. W., Lettenmaier, D. P., & Stouffer, R. J. (2008). Climate change. Stationarity is dead: Whither water management? *Science, 319*(5863), 573–574. https://doi.org/10.1126/science.1151915

Moser, S. C., & Ekstrom, J. A. (2010). A framework to diagnose barriers to climate change adaptation. *PNAS, 107*(51), 22026–22031.

National Climate Commission. (2010). *Belgian national climate change adaptation strategy*. National Climate Commission.

Naess, L., Bang, G., Eriksen, S., & Vevatne, J. (2005). Institutional adaptation to climate change: Flood responses at the municipal level in Norway. *Global Environmental Change, 15*, 125–138.

Nagoda, S. (2015). New discourses but same old development approaches? Climate change adaptation policies, chronic food insecurity and development interventions in northwestern Nepal. *Global Environmental Change, 35*, 570–579. https://doi.org/10.1016/j.gloenvcha.2015.08.014

Nelson, D. R., Adger, W. N., & Brown, K. (2007). Adaptation to environmental change: Contributions of a resilience framework. *Annual Review of Environment and Resources, 32*(1), 395–419. https://doi.org/10.1146/annurev.energy.32.051807.090348

Nelson, D. R., & Finan, T. J. (2009). Praying for drought: Persistent vulnerability and the politics of patronage in Ceará, Northeast Brazil. *American Anthropologist, 111*(3), 302–316. https://doi.org/10.1111/j.1548-1433.2009.01134.x

Neslen, A. (2018). Balkan dam projects could result in loss of one in 10 European fish species. In *The guardian*, 18/4/2018. Retrieved February 7, 2020, from https://www.theguardian.com/environment/2018/apr/18/balkan-dam-projects-could-result-in-loss-of-one-in-10-european-fish-species

Newell, P. (2015). The politics of green transformations in capitalism. In I. Scoones, M. Leach, & P. Newell (Eds.), *The politics of green transformations* (pp. 68–85). Routledge.

NOAA. (n.d.). *DROUGHT: Monitoring economic, environmental, and social impacts*. Retrieved February 7, 2020, from https://www.ncdc.noaa.gov/news/drought-monitoring-economic-environmental-and-social-impacts

O'Brien, K., Eriksen, S., Nygaard, L. P., & Schjolden, A. N. E. (2007). Why different interpretations of vulnerability matter in climate change discourses. *Climate Policy, 7*(1), 73–88. https://doi.org/10.1080/14693062.2007.9685639

O'Brien, K. (2009). Do values subjectively define the limits to climate change adaptation? In W. N. Adger, I. Lorenzoni, & K. L. O'Brien (Eds.), *Adapting to climate change* (Vol. 56, pp. 164–180). Cambridge University Press.

OECD. (2013). *Water and climate change adaptation. Policies to navigate uncharted waters*. IWA Publishing.

Olsthoorn, A. A., & Tol, R. S. J. (2001). *Floods, flood management and climate change in the Netherlands*. Institute for Environmental Studies.

Oppermann, E. (2011). The discourse of adaptation to climate change and the UK climate impacts Programme: De-scribing the problematization of adaptation. *Climate and Development, 3*(1), 71–85. https://doi.org/10.3763/cdev.2010.0061

Palmer, M. A., Reidy Liermann, C. A., Nilsson, C., Flörke, M., Alcamo, J., Lake, P. S., & Bond, N. (2008). Climate change and the world's river basins: Anticipating management options. *Frontiers in Ecology and the Environment, 6*(2), 81–89. https://doi.org/10.1890/060148

Pahl-Wostl, C. (2007). Transition towards adaptive management of water facing climate and global change. *Water Resources Management, 21*(1), 49–62.

Pahl-Wostl, C. (2009). A conceptual framework for analysing adaptive capacity and multi-level learning processes in resource governance regimes. *Global Environmental Change, 19*, 354–365.

Pahl-Wostl, C. (2015). *Water governance in the face of global change*. Springer International Publishing.

Pelling, M. (2011). *Adaptation to climate change. From resilience to transformation*. Routledge.

Pelling, M., O'Brien, K., & Matyas, D. (2015). Adaptation and transformation. *Climatic Change, 133*(1), 113–127. https://doi.org/10.1007/s10584-014-1303-0

Perini, K. (2017). Green and blue infrastructure in cities. In K. Perini & P. Sabbion (Eds.), *Urban sustainability and river restoration* (Vol. 14, pp. 1–9). John Wiley & Sons Ltd.

Pierre, J., & Peters, B. (2005). *Governing complex societies: Trajectories and scenarios*. Palgrave Macmillan.

Resurreccion, B., Sajor, E. E., & Fajber, E. (2008). *Climate adaptation in Asia: Knowledge gaps and research issues in South East Asia*. ISET-International and ISET-Nepal.

Rijkswaterstaat. (2016). *Dutch water programme room for the river*. Retrieved February 5, 2020, from https://issuu.com/ruimtevoorderivier/docs/factsheet_dutch_water_programme_uk__47425562a3293d?workerAddress=ec2-54-89-43-103.compute-1.amazonaws.com

Scoones, I., Leach, M., & Newell, P. (2015). *The politics of green transformations*. Routledge.

Smit, B., Burton, I., Klein, R. J. T., & Wandel, J. (2000). An anatomy of adaptation to climate change and variability. *Climatic Change, 45*(1), 223–251. https://doi.org/10.1023/A:1005661622966

Smit, B., & Pilifosova, O. (2001). Adaptation to climate change in the context of sustainable development and equity. In J. J. McCarthy, O. Canziani, N. A. Leary, D. J. Dokken, & K. S. White (Eds.), *Climate change 2001: Impacts, adaptation and vulnerability* (pp. 877–912). Cambridge University Press.

Smit, B., & Wandel, J. (2006). Adaptation, adaptive capacity and vulnerability. *Global Environmental Change, 16*, 282–292.

Steffen, W., Grinevald, J., Crutzen, P., & McNeill, J. (2011). The Anthropocene: Conceptual and historical perspectives. *Philosophical Transactions Series A, Mathematical, Physical, and Engineering Sciences, 369*(1938), 842–867. https://doi.org/10.1098/rsta.2010.0327

Steffen, W., Rockström, J., Richardson, K., Lenton, T., Folke, C., Liverman, D., et al. (2018). Trajectories of the earth system in the Anthropocene. *Proceedings of the National Academy of Sciences of the United States of America, 115*(33), 8252–8259. https://doi.org/10.1073/pnas.1810141115

Stern, N. (2006). *Stern review report on the economics of climate change*. HMSO.

Swart, R., Robinson, J., & Cohen, S. (2003). Climate change and sustainable development: Expanding the options. *Climate Policy, 3*(1), S19–S40.

Tanner, T., & Allouche, J. (2011). Towards a new political economy of climate change and development. *IDS Bulletin, 42*(3), 1–14. https://doi.org/10.1111/j.1759-5436.2011.00217.x
Tanner, T., & Mitchell, T. (2008). Entrenchment of enhancement: Could climate change adaptation help to reduce chronic poverty? *IDS Bulletin, 39*(4), 6–15.
The Economics of Ecosystems & Biodiversity. (2013). *Glossary of terms*. Retrieved February 5, 2020, from http://www.teebweb.org/resources/glossary-of-terms/
The Editors of Encyclopaedia Britannica. (2020). *Hurricane Katrina*. Retrieved April 2, 2020, from https://www.britannica.com/event/Hurricane-Katrina
Thieken, A. H., Kienzler, S., Kreibich, H., Kuhlicke, C., Kunz, M., & Mühr, B. et al. (2016). Review of the flood risk management system in Germany after the major flood in 2013. *E&S,* 21 (2). https://doi.org/10.5751/ES-08547-210251.
Trottier, J. (2004). Introduction. In J. Trottier & P. Slack (Eds.), *Managing water resources past and present* (pp. 1–12). Oxford University Press.
UBA. (2013). *Water Resource Management in Germany: Part 1, Fundamentals: Umweltbundesamt of Germany.*
UBA. (2016). *AMICA–adaptation and mitigation–an integrated climate policy approach*. Retrieved February 25, 2020, from https://www.umweltbundesamt.de/en/topics/climate-energy/climate-change-adaptation/adaptation-tools/project-catalog/amica-adaptation-mitigation-an-integrated-climate
UNESCO. (2009). *Introduction to the IWRM guidelines at River Basin level*. UNESCO.
United Nations. (2015). *Paris agreement*. Retrieved January 3, 2020, from https://unfccc.int/sites/default/files/mphasi_paris_agreement.pdf
US Fish and Wildlife Service. (2019). *About the WSR Act*. Retrieved February 26, 2020, from https://www.rivers.gov/wsr-act.php
WDR. (2003). *World disaster report: Focus on ethics in aid*. International Federation of Red Cross and Red Crescent Societies.
Wilby, R. L., & Dessai, S. (2010). Robust adaptation to climate change. *Weather, 65*(7), 180–185. https://doi.org/10.1002/wea.543
World Bank. (1993). *Water resources management: A world bank policy paper*. The World Bank.
World Meteorological Organization. (2021). *2020 was one of three warmest years on record*. Retrieved December 27, 2021, from https://public.wmo.int/en/media/press-release/2020-was-one-of-three-warmest-years-record

4

Climate Change Adaptation and Water Management in South Korea

4.1 Climate Change in South Korea

4.1.1 Scientific Evidence of Climate Change in South Korea

Scientists confirm that the climate is rapidly changing in South Korea (Korea Meteorological Administration, 2014). The annual average temperature in Korea in the twentieth century increased more rapidly than the global average temperature, increasing by 1.2 °C between 1981 and 2010 (Korea Meteorological Administration, 2017). This temperature is 0.8 °C higher than the global average temperature rise, which showed a 0.4 °C difference in the same period (NOAA National Centers for Environmental Information, 2019). Temperature change, moreover, is accelerating. While the annual average temperature increased by 0.23 °C per decade between 1954 and 1999, it increased by 0.41 °C per decade between 1981 and 2010, and from 2001 to 2010 by 0.5 °C. The number of extremely low-temperature cases decreased, while extreme hot temperature days increased after the late 1980s (Korea Meteorological Administration, 2014).

Y. h. Kang, *Climate Change Adaptation in River Management*, Palgrave Studies in Water Governance: Policy and Practice,
https://doi.org/10.1007/978-3-031-10486-2_4

Precipitation patterns have changed as well. As of 2014, the average annual precipitation (1412 mm) increased by 7.4% compared to the average precipitation from 1980 to 2010 (1315 mm). Particularly, precipitation in summer, which is influenced by the monsoon, has increased significantly since the late 1990s. This increase can be mainly attributed to more frequent heavy rainfall during the typhoons passing through South Korea (Korea Meteorological Administration, 2014).

Climate change is projected to continue in the future. Although different future scenarios have different projections regarding expected temperature changes for the Korean Peninsula, it is generally agreed by scientists that the warming tendency is very likely to continue at a faster rate than the global average (Korea Meteorological Administration, 2017). Annual precipitation could increase by 17% in the twenty-first century in the Korean Peninsula, according to the two scenarios developed by the Korea Meteorological Administration and the National Institute of Meteorological Sciences (RCP2.6 and RCP6.0). At the same time, the period without precipitation will be extended, and this could lead to drought. The number of heavy rainfall days and the intensity of precipitation are also likely to increase (Korea Meteorological Administration, 2017).

4.1.2 Impact on Water Management

The impact of climate change on Korean society is already visible. Temperature and precipitation changes have significant effects on water management, forestry, agriculture, fisheries, industry, energy, health care, and housing (Government of Korea, 2015). South Korea's changing social structure, characterised by its ageing population and income polarisation, may increase social vulnerability to climate change. This section briefly describes how climate change can affect society with a focus on water management.

The increase in rainfall variability and temperature are likely to alter runoff variability in the river basins (Chang et al., 2007). The seasonal differences in precipitation could also increase, leading to both more flood and drought risks (Government of Korea, 2015). In addition, more frequent and intensive typhoons and accompanying floods can result in severe damage (Climate Change Policy Research Institute, 2012). Between 1990 and 2008, floods accounted for the highest number of victims among natural disasters in South Korea (Climate Change Policy Research Institute, 2012). Economic loss is another problem. In 2003, property damage caused by Typhoon Maemi was approximately 4590 million USD. Typhoon Ewiniar in 2006 caused 380 million USD worth of damage (Kang et al., 2013).

Although the population density is highest in Seoul and its surrounding area (i.e., Gyeonggi province), these areas are less vulnerable to floods than other regions in South Korea. This phenomenon can be explained by the difference in the economic capacity of the sub-national governments. Seoul and Gyeonggi province are equipped with better structural and non-structural measures for flood prevention compared to other regions (Kim et al., 2012).

Another impact of the changing precipitation is the increasing risk of drought. Since 1966, water demand in South Korea has increased; because of its growing population together with industrial development and per capita, water use has increased in all regions except Seoul. Climate change is causing additional water stress on top of other existing problems (Chang et al., 2007).

4.2 Adaptation Policy

4.2.1 National Climate Change Adaptation Policy

In Korean environmental policy, the concept of climate change adaptation first appeared in the third Climate Change Master Plan formulated for 2005–2007. The detailed policy was, however, decided when the first National Climate Change Adaptation Master Plan was launched in 2008 (Ministry of Environment, 2008). In 2010, the first National Climate Change Adaptation Policy (NCCAP) with legal force was established for 2011–2015 in compliance with the Low Carbon Green Growth Act (Government of Korea, 2010). The NCCAP is the master plan for national and local governments' action plans and is to be reviewed and subsequently revised every five years. The second NCCAP was launched in 2015.[1] Twenty ministries and governmental agencies were involved in the planning process of the NCCAP, including the Ministry of Environment, Ministry of Agriculture and Food, the Ministry of Health, and the Ministry of Land, Infrastructure and Transport (MOLIT)[2] (Government of Korea, 2015).

The interviewees for this research evaluated the current adaptation policy in South Korea as "systematic", pointing out that adaptation plans have been established from the national to local level consecutively. All sub-national governments are obliged to set up an adaptation plan by law. Consequently, all 17 provincial governments and 226 local governments have completed their action plans for climate change adaptation.

> By law, local governments should make adaptation plans/strategies and the plans were established in a top-down way. This approach has advantages and disadvantages. The Korean case is received as a very successful case in other countries because it is a rare case that a country has an adaptation plan from the national to the local level, not leaving a single local government out. (Anonymous, from Korea Environment Institute, personal communication, 2 March 2017)

Nevertheless, the top-down order to develop action plans did not lead to the active implementation of the plans. One interviewee pointed out that local governments set up adaptation plans because it was compulsory, but the plans are often not implemented in practice due to lack of awareness and resources. This results in too big a gap between planning and implementation (Anonymous 1 at Gyeonggi Research Institute, personal communication, 13 March 2017). The institutional background of the

---

[1] The third NCCAP was announced in 2020 for 2021–2025, but this policy is beyond this book's scope as the data collection for this book was conducted between 2015 and 2019.

[2] The name of the ministry has changed several times since it was created. The Ministry of Construction was established in 1961. It became the Ministry of Construction and Transport in 1994. President Lee, Myung-bak changed the ministry again into the Ministry of Land, Transport and Maritime Affairs in 2008. Then in 2013, part of the ministry's task was separated from it, and it became the Ministry of Land, Infrastructure and Transport (MOLIT). In this dissertation, the ministry is referred to as the Ministry of Land, Infrastructure and Transport.

top-down decision-making process in South Korea will be further explained in Sect. 4.5.2, which addresses the strong presidency within Korean politics.

Another reason behind ineffective implementation pointed out during interviews was the stakeholders' low awareness of climate change adaptation in local communities. One interviewee said that adaptation remains as a policy concept discussed only by planners such as public officers and researchers.

> Compared to other developed countries, the government is not lagging. However, if we talk about substantiality, adaptation is still in the hands of government officers and researchers. (Ahn, B., personal communication, 28 February 2017)

Weak cooperation among ministries is also referred to as a limitation of the current adaptation policy. The second National Adaptation Plan is a compilation of the policies from different ministries. The Minister of Environment is the Chair of the Climate Change Adaptation Council, and the council members are the senior managers of each ministry. The detailed planning and implementation of adaptation policy measures are undertaken by each ministry (Government of Korea, 2015). Although the Ministry of Environment coordinates overall adaptation policy, including the National Adaptation Plan, it does not have the authority to take part in the policy creation of other ministries or governmental agencies. As a result, the national adaptation plan cannot function as a coordinating plan leaving the possibility of conflicts between different policies.

4.2.2 Low Carbon Green Growth Act

Although the Low Carbon Green Growth Act was not established for climate change adaptation, this act has heavily influenced the planning stage of the first adaptation policy in Korea. Notably, the Four Rivers Project, the main case in this book, was implemented as part of the green growth policy.

In 2008, the South Korean government announced the Low Carbon Green Growth Act, declaring the country's intention to become the forerunner in the global green economy. When President Lee Myung-bak participated in the G8 summit as a guest invitee, he was apparently impressed by other countries' initiatives to reduce greenhouse gas emissions (Byun, 2011). After returning from the summit, Lee proposed 'Low Carbon Green Growth' as a new national vision for the next 60 years during his televised speech on the 60th anniversary of National Foundation Day (Ministry of Culture, Sports and Tourism, 2008). Subsequently, the government vigorously pushed the green growth policy. The National Strategy for Green Growth and the first Five-Year Plan for Green Growth were established. The Five-Year plan aimed to allocate 107.4 trillion KRW (approximately 92.4 billion USD as of 2019) for the green growth plan between 2009 and 2013, amounting to 2% of GDP per year (Jones & Yoo, 2011).

There have been discussions on the term 'green economy' for more than two decades (e.g., Pearce et al., 1989; Ferguson, 2015; Brand, 2012); however, the green economy and the related concept of green growth have received little attention from policymakers until the 2008 global financial crisis (Ferguson, 2015). Among the green growth initiatives pursued by different countries at that time, South Korea's policy was notable because it focused not only on mitigation but also on adaptation. The first goal of the National Strategy for Green Growth was climate change adaptation. Water management was heavily weighted within the adaptation strategy and a large share of the adaptation policy budget was allocated to the Four Rivers Project (Jones & Yoo, 2011). The budget for 'strengthening adaptive capacity to climate change' was the largest in the Five-Year Plan for Green Growth, accounting for 34% of the total budget (31.4 billion USD out of 91.8 billion USD). Moreover, within this category, the Four Rivers Project held the biggest portion (Jones & Yoo, 2011).

Another feature of the Korean green growth policy was the strong drive for a 'green new deal' which was part of the Five-Year Plan. Designed as a stimulus plan, the goal of the Green New Deal policy was to stabilise the macroeconomy by creating jobs. About 44 billion USD was allocated for the Green New Deal policy between 2009 and 2012 and the Four

Rivers Project had by far the largest budget (36.8%) within the Green New Deal policy (Government of Korea, 2009; Yun, 2010).

The design of the Low Carbon Green Growth policy indicates that the focus on economic interest was very strong, giving the impression that environmental protection is a subordinate goal of the policy. Yun (2010) criticises the policy in her article "Not so green: a critique of South Korea's growth strategy" pointing out that it is based on a growth-biased green discourse.

The decision-making process of the Low Carbon Green Growth policy shows the strength of South Korea's presidential institution. After the president's announcement of the policy, every ministry in South Korea developed plans and initiatives using the 'green' label (Yun, 2010), while social discussion and consensus were lacking in the legislation process (Youn et al., 2010). The Four Rivers Project which was part of the Green Growth policy faced similar criticism during its planning and implementation process as can be seen in the following section.

4.2.3 Four Rivers Project

The four major rivers in South Korea refer to the Han, Nakdong, Geum, and Yeongsan. The basins of the four rivers account for approximately 64% of South Korea's land. Therefore, managing the river basin areas has a significant influence on the landscape. Since the 1960s when industrialisation started in South Korea, the major rivers have been the subject of land development. River engineering technologies including large dams and dredging were adopted for river management (see Sect. 4.3.2 for more details). The Four Rivers Project was in line with the historic management of water in South Korea.

The Four Rivers Project, announced in 2008, was implemented as South Korea's leading climate change adaptation scheme within the Low Carbon Green Growth policy. The initial budget was 19.1 billion USD but the total spending as of 2019 was over 20 billion USD (Ministry of Culture, Sports and Tourism, 2009a; Board of Audit and Inspection, 2018). It was an ambitious project primarily consisting of building 16 weirs on the major rivers and 2 new dams on their tributaries, reinforcing banks, dredging the riverbeds, aligning waterways, and enlarging 96 agricultural reservoirs (Ministry of Culture, Sports and Tourism, 2009a). The MOLIT and its sub-agency, the Korea Water Resources Corporation (K-water) directed the entire project process, which was implemented within three years between 2009 and 2011.

The Four Rivers Project had four goals: (i) adaptation to climate change, (ii) coexistence of nature and human beings, (iii) re-creation of land, and (iv) establishment of the basis for balanced development and green growth (Ministry of Land, Transport and Maritime Affairs, 2009). The government under Lee Myung-bak argued that the project could contribute to not only adapting to climate change but also tackling the financial crisis and securing new engines for economic growth. In the announcement of the project, the government estimated that 340 thousand jobs would be created through the project. Below are the five critical tasks of the Four Rivers Project (Ministry of Culture, Sports and Tourism, 2009a).

1) Build weirs, small and medium-sized dams, and irrigation reservoirs in order to secure abundant water resources (1.3 billion m^3), which can prevent water shortages resulting from climate change. The demand for water would exceed the supply by 0.8 billion m^3 in 2011 and 1 billion m^3 in 2016.

2) Build dams, dredge river deposit soil, and strengthen old dikes, in order to prepare for floods bigger than 200-year floods, which will occur more often.

3) Restore the ecosystem by making ecological rivers and wetlands, and removing farming areas. Increase wastewater treatment facilities and algae removal facilities. By doing this, the water quality in major rivers will be improved.

4) Create bike lanes, walking paths, and sports facilities in order to make riverside areas into multipurpose spaces for lifestyle, recreation, tourism, culture, and green growth.

5) Promote various subsidiary projects which utilise the infrastructure and landscape created by the Four Rivers Project, in order to contribute to regional developments.

The tasks within the Four Rivers Project were criticised for conflicting with each other. Dam construction and dredging plans were placed together with ecosystem restoration. Journalists and NGOs also pointed out that river and wetland restoration was not planned in detail, unlike other tasks (Hong, 2009; KFEM, 2009).

In order to understand the background of the Four Rivers Project, it is useful to look into the Grand Canal Plan, which was Lee Myung-bak's major pledge during the 2007 presidential election. The main idea of the Grand Canal Plan was to widen, deepen and connect the Han River and the Nakdong River for shipping. As the candidate of Hannara Party, the major conservative party of South Korea at that time, Lee Myung-bak argued that the Grand Canal Plan would contribute to economic growth. Before the election, he visited the Rhine–Main–Danube canal in Nuremberg, Germany, with journalists and argued that canals can lead to economic improvement like in Germany (Hwang, 2006a). His Grand Canal Plan was controversially discussed during the election campaign because of its feasibility as well as the potential impact on the ecosystem. Nevertheless, once elected Lee pushed forward with his pledge (Lee, 2010).

Just two months after Lee's inauguration, he was confronted with massive protests. The protest was first triggered by the government's decision to resume importing US beef despite public concerns about 'mad cow disease (bovine spongiform encephalopathy)'. During the many demonstrations, protesters extended their agenda and began to oppose the Grand Canal Plan. The protests took place in the central areas of Seoul almost every evening from April to August 2008. Lee's approval rate dropped by more than 30% during this period (Korea Gallup, 2008a, b). Under political pressure, Lee announced the cancellation of the Grand Canal Plan.

The Four Rivers Project was announced about six months after the cancellation of the Grand Canal Plan as a part of the Green New Deal policy. Although the government denied the connection between the Four Rivers Project and the Grand Canal Plan (Office of the Prime Minister's Secretariat, 2009), it was revealed that the politicians who worked closely with Lee suggested the Four Rivers Project as an alternative project to the Grand Canal Plan (Kim, 2017; Park, 2010).

As justification for the massive investment in the Four Rivers Project, the government of Lee Myung-bak explained the necessity of the Four Rivers Project by arguing that damage from floods stood at 3.8 billion USD on an annual average (Ministry of Culture, Sports and Tourism, 2009a). Another primary rationale for the project was the water shortages which could be expected due to climate change impacts (Park, 2010). Therefore, the project was framed as "a precautionary solution for preventing floods and droughts" in the face of climate change (Ministry of Culture, Sports and Tourism, 2009a). The government also argued that the ecosystem could be protected through dredging because the rivers would not dry out during the winter dry season as well by preventing the inflow of pollution sources (Ministry of Land, Transport and Maritime Affairs, 2009).

Notably, the Four Rivers Project was planned before the first NCCAP. The NCCAP was therefore planned based on the assumption that the Four Rivers Project would be complete. In the action plan of the NCCAP, the Four Rivers Project was included under the task title "Maximisation of adaptation capacity of rivers" (Government of Korea, 2010).

The project was undertaken in haste (Kim, 2010a). The government skipped the preliminary feasibility study for the Four Rivers Project because the project was defined as an 'urgent disaster risk reduction measure' by the government (Board of Audit and Inspection, 2018). The archaeological field survey for cultural heritage in all of the project sites was completed within two months, and the environmental impact assessment for the project sites was done within four months. The public hearing took two months. The Ministry of Defence created an engineering unit mobilised for supporting the project implementation (Kim, 2010b).

As a result, the Four Rivers Project became one of the most controversial policies undertaken during Lee's term, which lasted until 2012. The environmental impacts (Park, 2009), procedural legitimacy (Kim, 2010a, 2011; Yun & Lee, 2010), and cost-effectiveness of the project were questioned by scholars from different disciplines. Environmental NGOs including the Korean Federation for Environmental Movements (KFEM) and scientists from different disciplines organised campaigns against the project. According to a public-opinion poll conducted by the National

Assembly Forum on Climate Change, most of the respondents evaluated the project negatively (Climate Change Policy Research Institute, 2012). The project was the subject of special inspections by the Board of Audit and Inspection four times between 2011 and 2018. The project was first positively reviewed when Lee was the sitting president, but the inspections undertaken after his term concluded negatively on the effects of the project (Board of Audit and Inspection, 2018; Song, 2011).

4.3 Water Policy

4.3.1 Water Quality Management and Ecosystem Conservation

Korean water quality policy began to be developed in the 1970s. While the initial focus was on drinking water provision and hygiene in urban areas, concerns about water quality emerged as water quality was worsening due to industrialisation, urbanisation, and lack of control (Ministry of Environment, 2010). According to the database of the Ministry of Health and the Seoul Metropolitan Government, the water quality of the Han River suffered severely from pollution in the 1970s (Jung et al., 2014). In 1989, a journalist reported that the Ministry of Construction and Transport found most tap water not suitable for drinking as it contained heavy metals, colon bacillus, bacteria, and ammonia nitrogen (Lee, 2006).

Water quality policy was substantially strengthened after several environmental accidents that occurred in the early 1990s. In 1991, 30 tonnes of phenol was leaked from an electronics factory to the Nakdong River. Again in 1994, organic solvent leakage occurred in the same river. Those chemical accidents gained public attention and increased concerns about environmental problems (Ministry of Environment, 2010). The then-president Kim Young-sam took the concerns of the residents along the Nakdong River seriously, mainly because the region was where his hometown was situated (Jung et al., 2014).

Consequently, the second Clean Water Provision Master Plan was implemented between 1993 and 1997, with a budget of 69 billion USD aiming to improve water quality in the four major rivers. Following this plan, the additional budget was allocated for wastewater treatment and advanced water treatment facilities (Jung et al., 2014). By the mid-2000s, water quality management in the four rivers had stabilised. Nevertheless, water quality regulations were often opposed by the residents in the upper stream areas of the rivers because development projects had to be restricted in those areas (Lee, 2012).

River basin-based management and integrated water management started to be discussed in the late 1990s. In 1999, a consultative committee for the Han River was formed comprising the residents around the Paldang Lake (the upper stream of the Han River), environmental NGOs, local governments, and research institutes. The committee established a water quality management master plan for the river basin and this model was spread to other river basin areas. In 2002, special acts for the Nakdong, Geum, and Yeongsan River basins were established through consultations with the stakeholders following the Han River model. Jung et al. (2014) state that this transition to river basin management policy shows the maturity of Korean democracy.

Concerns on ecosystem protection also began in the 2000s. In 2005, the Water Quality Environmental Protection Act was changed into the Water Quality and Water Ecosystem Protection Act. This amendment attempted a paradigm change to comprehensive hydro environment protection and management. The Basic Plan for Water Environment Management for 2006–2015 was also established. This plan aimed to restore 25% of the rivers of South Korea and to create riverine eco-belts in upper stream river basins. Due to the Four Rivers Project, however, the restoration plans based on the ecosystem conservation paradigm were not realised in practice (Jung et al., 2014).

4.3.2 Water Quantity Management

Besides water quality policies, water quantity policies have been planned and implemented. While the annual average precipitation in South Korea is 1283 mm, which is 140% more than the global average (Jung et al., 2014), the precipitation is concentrated during the summer season and there is a substantial difference between winter and summer precipitation. This meteorological characteristic has been the crucial cause for the majority of dam construction projects in Korea. The MOLIT has been the main ministry for handling water quantity issues.

The Korean government has been very active in building dams for controlling floods, securing water, and generating hydropower. As of 2010, there are 1302 large dams in South Korea and most of them (about 1182 dams) are for irrigation (Kang, 2012). According to the International Commission on Large Dams (2019), Korea is ranked sixth in the world by the number of its dams, following China, the United States, India, Japan, and Brazil.

Modern dams were first built in Korea during the Japanese occupation in the early twentieth century. At that time, Japanese companies led hydroelectric dam constructions in Korea in order to develop and test their machinery (Kang, 2012). As most of the dams built by the Japanese were in the territory of North Korea, the South Korean government made efforts to build hydroelectric dams after Korea was divided into two (Kang, 2012). The 10-Year Plan for Comprehensive Water Resource Development was established in 1965, aiming fast development of dams and estuary barriers. The Act on Specific Multi-purpose Dams was legislated in 1966 and the public enterprise, the Korea Water Resources Corporation (K-water), was established in 1967 as the governmental agency for developing and managing water resources. With this law and the dedicated agency for water resource development, multi-purpose dam construction was accelerated after the late 1960s (Kang, 2012; Lee, 2005). Besides K-water, the MOLIT (K-water's superior authority) and construction companies which were awarded the contracts strongly drove the dam development.

The pace of the dam development policy was slowed down in the late 1990s since public concerns on the environment rose. After floods in 1998 and 1999 caused more than 800 million USD in damage, the MOLIT set up the Integrated River Basin Flood Control Plan (MOLIT, 2012a). The plan was the first attempt to utilise non-structural measures and green infrastructure for controlling floods. In 1999, the Act on Specific Multi-purpose Dams was abolished and substituted with the Act on Construction of Dams and Assistance to Their Environment. This revision came in response to increasing problems relating to dam construction projects, such as large-scale emigration from submerged areas.

Nevertheless, the MOLIT and the K-water have continued dam development. The MOLIT established the Long-Term Plan for Dam Construction (2012–2021) and indicated its plan to build four large size dams as well as eight middle and small size dams with a budget of approximately 3 billion USD, primarily for water security (MOLIT, 2012b). What is different from their past policy is that social consultation on new dam projects was given more emphasis during the planning process. Besides the development projects pursued by the MOLIT, the Ministry of the Interior and Safety has implemented a contest among local governments called 'Making Small Streams Beautiful and Safe'. This contest aimed to encourage local governments to develop streams and small rivers (Ministry of the Interior and Safety, 2017). The second National Climate Change Adaptation Plan included those policies (Government of Korea, 2015).

4.3.3 Critical Junctures and Path Dependency

The previous sections on water quality and quantity management discussed the water policies that affected the Four Rivers Project with their brief historical background. While climate change adaptation is a new policy agenda in South Korea and major changes from the first to the third NCCAP cannot be observed, water management policy has a relatively long history and it went through some changes. This section focuses on grievous events that influenced the institutions of water quality and quantity management in Korea. The main database used here is news articles found in two major newspapers (*Kyunghyang Shinmun* and *Dong-a Ilbo*).

Several critical moments can be found in Korean water management policy between 1961 and 2019. Almost every year until the 1980s, hundreds of people were killed by floods or landslides due to heavy precipitation. Most floods in South Korea occur due to typhoons and heavy precipitation in the summer season (K-water, 2015; Han River Flood Control Office, 2010). In response to severe flooding, river engineering methods were promoted by the media, politicians, and the public (Kang,

1976). As some flood damage reduction could be observed after dam completion, the media praised dam construction projects. A newspaper article states that "flood control transformed the map of Korea" (Lee, 1979).

Criticism against water resource development projects appeared for the first time in the early 1980s ("The negative impact", 1983). In 1981, an association for nature conservation demanded an environmental impact assessment be conducted before construction was to begin on an estuary dam of the Nakdong River ("Environmental conservation", 1981). Yet, environmentalists hesitated to oppose the dam construction because they were afraid to be put on the government blacklist (Lee, 1984). The public opinion on dam construction was reflected in the news articles that reported floods in the 1980s arguing the necessity of dam construction and land development projects (Kim, 1986; Choi, 1970).

Chemical accidents in the early 1990s provided windows of opportunity to strengthen water quality policy. As mentioned in Sect. 4.3.1, the phenol contamination incident in 1991 and the waste organic solvent leakage in 1994 led to institutional changes. In 1994, the tasks of drinking water provision and wastewater treatment were transferred from the Ministry of Construction (former name of the MOLIT) and the Ministry of Health to the Environmental Agency, which was raised to the Ministry of Environment in the same year (Jung et al., 2014; Ministry of Environment, 2010). Also, a series of water quality regulations were established: The Clean Water Provision Master Plan (1993), Water Quality Management Plan (1994), and Water Management Master Plan (1996). The two chemical accidents are therefore regarded as significant critical junctures not only for water management policy but also for the general environmental policy of South Korea (Jung et al., 2014; Kim, 2013).

The negative aspects of dam construction started to be discussed more actively around 1990. In that year, the residents of Gangwon province refused dam development along the Han River, arguing that the upper stream areas (Gangwon Province) would have a higher flood risk instead of the downstream areas (Seoul and Gyeonggi Province) (Choi, 1990). In the late 1990s, controversy over the Youngwol Dam broke out. The dam construction was planned in the upper stream of the Han River for flood control. While the MOLIT and K-water strongly pushed the dam construction plan, environmental NGOs including the KFEM organised a campaign against the dam development project which was planned to be built in an ecologically rich area (Hwang, 2006b). The anti-movement emphasised the beauty of the river landscape nearby the dam site, using the rare species in Youngwol as the symbol of their campaign. This strategy worked effectively on the broader public and the number of tourists to the planned submergence site increased (Ku, 2011). After three years of discussion and conflicts, the Youngwol Dam plan was finally cancelled on World Environment Day in 2000 (Lee, 2014). This shows well the significant turn in public awareness of environmental protection.

Despite these changes, the primary response to flood defence remained hard infrastructure development as stated in Sect. 4.3.2. The MOLIT typically announced new dam development plans after every big flood (Lee, 1999; Kim & Kim, 2006), and the arguments against large dams were described as NIMBYism (Chang, 1996). A mayor of Paju, a city which was flooded almost every year, said that the problem is the "natural state of the rivers" which needed to be fixed urgently (Gwon, 1998).

Table 4.1 presents a chronological list of critical events and significant institutional changes in the water management sector of Korea. As the table indicates, the Four Rivers Project which consisted of dam construction and dredging can be explained by the continuity of flood control policy since the 1960s. While water quality regulations have changed since the first pollution prevention regulation was established in the 1960s (Jung et al., 2014), measures for water quantity management have been fixed to grey infrastructure and engineering-based solutions represented by dams.

The consequence of the Four Rivers Project, however, provided an opportunity to change water policy. Serious algal blooms appeared in the major rivers covered by the project, triggering public controversy. Environmentalists and politicians from the opposition party (Democratic Party of Korea) described the algal blooms as "Green Tea Latte", referring to the thick, green beverage (Kim, 2012). Media attention on algae bloom and its impact on ecosystem as well as public health grew over

Table 4.1 Critical events and major policy/institutional changes in Korean water management between 1961 and 2019. Critical events are italicised and policy/institutional changes are underlined

	Quantity management By MOLIT	Quality management and ecosystem conservation By Ministry of Environment
1961	The first multi-purpose dam (Seomjingang-dam) constructed	
1972	*Severe flood along the Han River basin* Plan to construct 12 multi-purpose dams established	
1991		*Phenol contamination incident in the Nakdong River*
1993		Clean Water Provision Master Plan
1994		*Waste organic solvent contamination in the Nakdong River* Ministry of Environment established Water Quality Management Plan
1998	*Floods in Seoul and nearby provinces*	
1999	*Floods due to Typhoon Olga* Integrated River Basin Flood Control Plan	
2000	Cancellation of the Youngwol dam construction plan	
2001	1st Long-Term Plan for Dam Construction	
2009–2012	*Four Rivers Project*	
2012–2021	2nd Long-Term Plan for Dam Construction	
2017	*The opposition party's candidate (Moon Jae-in) was elected in the presidential election*	
2018	Announcement on stopping major dam construction Framework Act on Water Management	

time. The government tried to reduce the phosphorus concentration through chemical treatment, but this failed to solve the problem (Kim, 2018a).

After President Moon Jae-in from the Democratic Party of Korea seized power in 2017, the government responded to the criticism aimed at the Four Rivers Project. The new government decided to evaluate the effectiveness of the dams built during the project and demolish them if necessary. In 2020, a task force under the Ministry of Environment assessed the feasibility of different options and their impacts on water quality, ecosystem, and groundwater (Ministry of Environment, 2020) and decided to open some dams' gates and monitor changes (In Hyunwoo, 2021). However, no plan for dam removal was decided under Moon's government.

Furthermore, heavy criticism over the results of the Four Rivers Project fuelled again the discussion on integrated water resource management among policymakers, researchers, and environmentalists (personal communication, 3 March 2017; 17 March 2017). Moon Jae-in's government initiated a reform in water policy with a focus on the integration of water management. Unlike similar attempts undertaken by previous governments, the government's initiative gained political support from stakeholders due to the generally negative opinion of the Four Rivers Project. In June 2019, the Framework Act on Water Management took effect intending to integrate water management institutions. The major change brought by the new law was that tasks of water quantity management were transferred from the MOLIT to the Ministry of Environment. The government highlighted that there would be no more large dam construction projects undertaken in the future (Kim, 2018b).

Nevertheless, environmental NGOs including the KFEM criticised the reform because river management tasks remained in the MOLIT. This seemed as the result of a political compromise between the ruling party (Democratic Party of Korea) which wanted a notable achievement even though it was partial and the opposition party (Liberty Korea Party) which had a close link to the MOLIT ("Summary", 2018). The conclusion of the water policy reform shows the influence of the MOLIT is still strong in Korean politics. In order to explain this phenomenon, social discourses and the formal and informal institutions of Korean politics are analysed in the following two sections.

4.4 Discourse Analysis

4.4.1 Developmentalism

Developmentalism can be defined in various ways. While developmentalism in economics refers to a model of economic development pursued in the 1950s–1960s, particularly in Latin America (Mainwaring, 1992; Sikkink, 1991), in this book, developmentalism refers to the social discourse which prefers economic development over environmental protection. This term is a direct translation from a Korean word " [*gaebal-jui*]" which is regarded as a prominent social discourse in South Korea (Cho, 2003, p. 135; Lee, 2005; Kim et al., 2015). The term indicates the belief that economic growth is the most important goal not only for the state but also for individuals. Developmentalism can be, therefore, also called growthism in the Korean context (Yoon, 2016; Yang, 2011).

Developmentalism is one of the most forceful social discourses in South Korea. As the Korean economy was underdeveloped after the Japanese colonial period (1910–1945) and heavily hit by the Korean War (1950–1953), a strong desire to be better off economically existed among both politicians and the public. Cho Myung-rae, a former Minister of Environment in South Korea, argues that a development-oriented paradigm has been dominant in South Korea and that this paradigm was re-enforced after democratisation in 1987 (Cho, 2006). He argues that, since the 1990s, developmentalism has been particularly realised through land development (Cho, 2003).

Developmentalism can be easily found in many governmental policies in South Korea. Large-scale land development projects led by the national government are the symbol of this discourse (Cho, 2003). Politicians often support large-scale urban development and infrastructure construction projects and make attractive pledges to win over voters during election campaigns (Jeon, 2011). Once elected, politicians carry out the development projects with the assumption that the voters will approve of the projects. Many interviewees framed the Four Rivers Project as a development project which resembled previous cases such as the Saemangeum Seawall Project or the Gyeongin Canal Project promoted by former political leaders. Completed in 2010, the Saemangeum Seawall is the world's longest dike with a length of 33 km. The project aimed to create farmland, an industrial complex, and a freshwater reservoir on the flat land created by the seawall (The Institute for Democracy, 2011). The Gyeongin Canal Project was first planned for flood protection in the Incheon area in the Han River basin, and the purpose of the project extended to building a waterway between Seoul and Incheon. Since the plan was announced in 1995, the project was the subject of controversy due to its environmental impact and economic feasibility. Nevertheless, the construction of the 18 km canal started in 2009 and was completed in 2012 (Park, 2017).

> Development projects are so attractive for politicians. Nobody failed to be elected in presidential elections with a pledge of a mega-scale project. Roh Taewoo was able to be elected as president [of South Korea] with his pledge on the Saemangeum Seawall Project and the KTX (express railway) construction plan. The Gyeongin Canal was a sub-national project, but (national level) politicians from all major parties supported it. (Hong J. H., personal communication, 21 March 2017)
>
> Every new government has planned and implemented large-scale civil engineering projects due to their obsession with an economic boom. The Saemangeum Seawall Project and the Four Rivers Project are such cases. (Ahn, B., personal communication, 28 February 2017)

The Four Rivers Project was pursued under the Green New Deal plan, which was part of the Green Growth policy. The aim of the Low Carbon Green Growth Act is "to utilise green technology and the green industry as the new driving force of economic growth … in order to achieve harmonious development of economy and the environment". It is evident in this statement that the policy's focus was economic growth. The name of the Green New Deal plan itself shows the aim of the scheme: economic stabilisation was at the centre of the scheme. Although the government framed the Four Rivers Project as a climate change adaptation policy, its economic benefit was emphasised frequently during the project implementation. President Lee referred to the Four Rivers Project as "a role model of green growth" (Newsis, 2009). The government advertised the Four Rivers Project as "the biggest development project by the public sector in Korean history" (Ministry of Culture, Sports and Tourism, 2009b).

In an opinion poll conducted in a town where one of the weirs of the Four Rivers Project was constructed, residents who supported the project replied that benefits for the local economy were the primary reason for their support. In other words, other aspects of the project such as the impacts on the ecosystem and the landscape change were not considered more important than the economic benefit from the project (Jeon, 2011). One interviewee from an environmental NGO pointed out that the economic growth paradigm from the fast growth period in the past has remained prevalent in sub-national-level governments:

> South Korea is still pursuing development and the Four Rivers Project was just the tip of the ice-berg. Even if we dismantle the social developmental state structure, it won't stop. There are similar projects pursued by local governments. Residents want their town to become like Seoul. The growth paradigm is still dominant. (Yeom, H., from KFEM., personal communication, 16 March 2017)

In summary, developmentalism can be referred to as one of the dominant social discourses in Korean society. Although environmentalism has grown substantially in South Korea, development projects are still welcomed by local government leaders. Cho (2006) argues that in Korea environmental protection is emphasised ostensibly, but development is encouraged in reality.

4.4.2 Techno-Centrism

Techno-centrism is a widespread discourse not just in South Korea but in modern society. The technocentric paradigm emerged with the Scientific Revolution in the seventeenth century. The underlying assumptions of the technocentric worldview can be summarised as the superiority of humankind to nature. The assumption of continued economic growth and technological innovation is strong (Gladwin et al., 1995).

The interviewees in South Korea pointed out that the technocentric worldview was widely spread in the country. This discourse can be considered as connected to the fast industrialisation of Korean society. Rapid economic growth has often been regarded as the country's success, and technological advancement has been highly valued as the primary engine for growth. Political leaders often emphasise that in such a disadvantaged country, without abundant natural resources like South Korea, technology and human resource development are extremely important for competing in the global economy (e.g., Kim, 2001).

A technocentric discourse can be found in environmental policies as well. Although techno-centrism is often placed in opposition to ecocentrism in a sustainable development paradigm spectrum (Gladwin et al., 1995; Brand, 2015), technology itself is not denied as a way to protect the environment by environmentalists. The central assumption of the ecological modernisation theory in the 1980s, for example, was that economic prosperity and environmental concerns are not in a zero-sum trade-off relationship (Jänicke, 1985; Huber, 2000). Joseph Huber (2000) suggests ecological modernisation as an advancement of the industrial society, acknowledging the role of technology.

Korean water policy has been under the influence of techno-centrism. Particularly, large dams have been the symbol of industrialisation and economic development. During the period of fast economic growth between the 1970s and 1980s, presidents came to dam completion ceremonies to praise economic development and technical advancement, which were represented by large dams (Jung, 1972; Paldang Dam Completion Ceremony, 1974; Choi, 1985). Canal construction in major rivers including the Han River was proposed by K-water, a sub-agency of the MOLIT, and this proposal appeared repeatedly in the media until the 1990s ("Natural disasters", 1962; Park, 1968; Kim, 1970; Choi, 1995). The proponents of canals argued that canals are the symbol of technological and economic advancement. The Grand Canal Plan, which was pushed forward before the Four Rivers Project, was in line with the long-standing proposal to promote canal construction.

Several interviewees of this study argued that the Four Rivers Project was also pursued following this technocentric attitude. One interviewee pointed out that the introduction of robot fish was promoted by President Lee Myung-bak during the Four Rivers Project as a water quality monitoring tool and this showed the worldview behind the project:

> In South Korea, there is a myth of the technological fix. For example, the robo–fish—it's not the robot fish itself, but what we need to focus on is the idea that 'we can solve anything with technology'. When environmental scientists brought up the issue that water quality would get worse after the project, the government and the supporters of the project argued that technology could fix the problem, saying that "Look, astronauts can drink water purified from their own pee. We have that kind of (high) technology". Behind such a statement, techno-centrism exists. (Jo, H., personal communication, 10 April 2017)

The "Smart River" concept suggested by the MOLIT in the first NCCAP Action Plan (Government of Korea, 2011, p. 316) is another example of techno-centrism in environmental policy. According to the first NCCAP Action Plan, Smart River is a river monitoring system based on information technology in order to optimise water quality and ecosystem management. In order to develop the monitoring system, software engine development and ubiquitous technology applications were planned in the Action Plan.

The technology-based approach can be found not only in the water sector. In March 2019, the ultra-fine particle concentration level in the air was at a record high for five days in a row. As public opinion worsened, the Minister of Environment announced its plan to install outdoor air purifiers (Oh, 2019). This episode depicts how dependent the Korean government is on technological solutions for environmental problems.

4.4.3 Environmental Awareness

Although developmentalism and techno-centrism are two dominant social discourses in South Korean environmental politics, environmental movements cannot be disregarded. From the 1960s to the 1970s, when industrialisation was heavily pursued, environmental movements were treated as an obstacle to economic development. Nevertheless, the Korean Pollution Research Institute, the first environmental NGO of South Korea, was established in 1982, reflecting growing public concerns with pollution. This organisation contributed to publicising health issues related to pollution. Over 500 residents in Onsan, a town in southeastern Korea nearby an industrial complex, reported various symptoms including neuralgia, skin irritation, nausea, and pruritus but the cause was unknown in the early 1980s (KFEM, 2000; Is the Onsan Illness False, 1985). The Korean Pollution Research Institute publicised their argument that it was a pollution disease due to the pollutant from heavy-chemical factories, sparking an environmental movement in South Korea (KFEM, 2000).

In 1993, the Korean Federation for Environmental Movements (KFEM) was organised by local environmental NGOs. Since then, the Korean environmental movement grew fast. While there were only 5 environmental NGOs officially registered with the government in 1987, 626 environmental NGOs were registered by 2013 (Jung et al., 2014; Ministry of Environment, 2013). The growth of environmental groups can be attributed to increasing awareness of environmental issues as well as democratisation after the June Democratic Uprising in 1987 (Ku, 2011; Kim, 2012). Kim (2001) states that the development of democracy after 1987 significantly contributes to solving many environmental issues. Indeed, several pollution accidents in the early 1990s such as the Phenol contamination incident and waste organic solvent contamination in the Nakdong River led to the establishment of the Ministry of Environment.

Escalated awareness of the environment among the public provided an impetus to overcome the conservative perspective of the budget department, the NIMBYism of residents, and the lobbying from industry. Opinion polls show that environmental awareness has gradually increased since the 1980s. In a survey undertaken in 1982, only 5.7% of the respondents chose environmental problems as the most critical social issue, while the percentage increased to 20.8% in 1990 and 33.1% in 1996 (Yang, 1993; Ku, 1999; Lee, 2011). When the Four Rivers Project was being implemented, public opinion against the project was not negligible. According to two surveys conducted at the early stage of the project, opposition to the project was stronger than support for it: 51% opposed versus 35.9% supportive of the project in December 2009 and 49.9% opposed versus 36.7% supportive in March 2010 (Realmeter, 2010a). Many respondents to the survey pointed to the Four Rivers Project as the main reason for the ruling party's defeat in the 2010 local elections (Realmeter, 2010b). In another survey conducted by Kim et al. (2015), environmental protection, procedural legitimacy, and high cost were most frequently pointed out as the causes of objection to the project.

Higher awareness of environmental issues is not always linked to there being a more active environmental movement. In a poll conducted in 2010 about the willingness to join a demonstration on environmental issues or to donate to an environmental NGO, the majority of the respondents gave negative answers (Lee, 2011). One interviewee mentioned that environmental problems such as climate change have a lower priority compared to other social problems on the Korean social policy agenda.

> If you ask people whether climate change is an important issue, about 95% would say yes. But if it is compared to other issues, it is only at the 8th place in terms of their priority. Unemployment or national security are considered 'more critical issues'. (Yun, S. J., personal communication, 6 March 2017)

Particularly, adaptation to climate change has not received much attention even from environmental NGOs in South Korea. Compared to other environmental issues such as energy transition and nature conservation, adaptation has remained low on the priority list on their agenda. Such a low level of interest in climate change adaptation led to the weak counter-discursive power of environmental NGOs when the Four Rivers Project was framed as an adaptation project by the government. In a lawsuit brought by major environmental NGOs against the government regarding the Four Rivers Project, the Supreme Court ruled that public interest (i.e., flood risk reduction) gained through the project was more significant than the ecosystem loss (Ahn, 2014). This ruling shows that there was no strong discursive power which could connect the policy subjects of adaptation and environmental protection.

4.5 Institutional Analysis

4.5.1 The Developmental State (as a Formal and Informal Institution)

The concept of the developmental state was created in order to explain the economic growth strategies pursued by East Asian governments after the Second World War (Johnson, 1987). Chalmers Johnson first explained Japanese economic policy with the developmental state model and then extended it to other countries, including South Korea (Johnson, 1982, 1987).

State-controlled finance is the most crucial aspect of the developmental state (Johnson, 1987). The developmental state intervenes in industry activities with various financial policies such as controlling tax credits, interest rates, exchange rates, and subsidies, aiming to achieve high economic growth through effective use of finance (Caldentey, 2008; Johnson, 1982). Johnson (1999) argues that the developmental state is not the same as an authoritarian state (authoritarianism) because the leaders of the developmental state derive legitimacy from their performance.

In South Korea, the developmental state institutions were created under the military regime in the 1960s. A robust economic bureaucracy, the Economic Planning Board, was established for the same function as the Japanese Ministry of Trade and Industry (Kim, 2008). In addition, the first Five-Year Economic Development Plan was instituted for promoting labour-intensive and light industries. The Korean government set up seven Five-Year Economic Development Plans in total until 1996, with shifting emphasis in terms of which key industries were to be promoted.

The essential instrument of the Korean government for economic development was financial control through nationalising the banking industry, allocating resources strategically to selected industries, and socialising investment risk (Kim, 2005). The government was able to intervene in the economy effectively because the banking industry was practically nationalised (Kim, 2008). This strategic investment guided by the government led to rapid industrialisation in South Korea.

The state-led industrialisation policy resulted in the "back-scratching" relationship between the government bureaucracy and private business (Johnson, 1982, pp. 309–310). The selected corporations received various support in the form of reduced taxes, export subsidies, and access to cheap credits from the government while being protected from global competition (Kim, 2005, 2008). Highly concentrated '*chaebol*' groups (family-controlled large conglomerates) were able to grow explosively through governmental support (Haggard & Kang, 2015). In turn, the *chaebol* groups followed the state's economic standard and played the role of delivering public goods and infrastructure required by the state (Kim, 2008).

The line between the public and the private was not clear. Woo-Cumings (1999) mentions that *chaebols* were regarded as "quasi-state organisations" while their profit remained private. During President Park Chung Hee's regime (1961–1980), for example, prominent business people joined monthly economic policy meetings with the president and top government officials. The policy consultation and coordination with the private sector was an "integral part of the policy formulation and implementation" (Kim, 2008, p. 16). In this process, a unique, consolidated state-business relationship was created in South Korea (Kim, 2005). This relationship was named the "developmental coalition" by Woo-Cumings (1999, p. 12). It consists of big businesses, political interest groups, and the state bureaucracy.

Ding (1994) argues that the distinctive features of state-society relations in East Asian countries did not disappear after industrialisation or democratisation. The blurred boundary exists between public and private, political and personal, formal and informal, official and nonofficial, governmental and market, legal and customary, and procedural and substantial (Ding, 1994). Ding's argument can be applied to the case of South Korea. The institutions of the developmental state have lasted even after the rapid economic growth period came to an end. They continued to exert influence in various policy areas including environmental policy.

Notably, the construction industry has been closely linked to developmental state institutions. When the development of land and urban areas was pursued vigorously in the 1970s, the leading governmental officers in charge of it were those who implemented post-war restoration in the Ministry of Construction (Cho, 2006). Infrastructure building works provided major corporations the opportunity to accumulate their capital and expand their business territory to other industry sectors (Cho, 2006). For this reason, the Korean economy is highly dependent on the construction sector as compared to other industrialised countries (OECD, 2018). As of 2002, the value of orders received by the construction sector was worth 19.8% of GDP (Cho, 2006, p. 28).

The formal developmental state institution is represented by the MOLIT. Established in 1961, the Ministry of Construction has been tasked with infrastructure construction and land use planning. The name of the ministry has changed a few times: Ministry of Construction and Transportation in 1994, Ministry of Land and Maritime Affairs in 2008, and the MOLIT in 2013. Nevertheless, the ministry's core task has been infrastructure building, and this has not changed.

Regardless of the ruling party, large-scale development projects have been the central pledge during presidential elections. Cho (2006) argues that such tendency results from strong developmental institutions which form the development coalition. The major actors of the developmental coalition in the construction industry include development-related research institutes (e.g., the Korea Research Institute for Human Settlements and the Korea Institute of Construction Technology), businesses (e.g., the Construction Association of Korea, real estate-related associations, and construction companies), and politicians (Cho, 2006). In big national development projects, public corporations under the MOLIT (i.e., the Korea Land & Housing Corporation, the Korea Expressway Corporation, and K-Water) play the role of brokers between the government and industry. Cho (2006) criticises that many development projects in recent days have not been pursued out of necessity, but rather for their own sake.

With the decentralisation since the late 1990s, development projects have been pushed by provincial and local governments. They did not have authority to pursue development projects during the military regime until 1987, but since then, the local governments can exercise their authority. The first elections since 1960 for local assemblies began in 1991 and for governors and mayors in 1995 (Jung et al., 2014). The governors and mayors elected in those local elections started to propose and implement major development projects in order to win the next election. Survey results show that residents have fully supported development projects, and local governments have been more interested in development than environmental protection (Jung et al., 2014, p. 52; Cho, 2006). Therefore, the local government can be regarded as part of the developmental coalition, as one interviewee pointed out.

> The MOLIT wouldn't oppose something financially beneficial for themselves. Also, those who benefit [from development projects] in the local regions, construction companies, civil engineers and researchers who have projects funded by the government, namely the stakeholders, make a coalition. Without accurate information, local residents follow the argument that development projects can create employment and other benefits for them. (Hong, J. H., personal communication, 21 March 2017)

Environmental protection was not prioritised in the pro-development policy agenda. The first environmental law in South Korea was the Air Pollution Prevention Act legislated in 1963; however, the law was not implemented until the enforcement ordinance was enacted in 1967 (Jung et al., 2014). The first governmental departments in charge of the environment were established in the Ministry of Health in 1967 and were called the Department of Environmental Hygiene and the Air Pollution Team. Nevertheless, discussions on environmental problems were taboo among governmental officials because pollution was regarded as the symbol of economic growth and prosperity. People in business opposed regulations on environmental protection, and politicians followed the business sector's interest. Although there were chemical accidents in the 1960s, they did not affect governmental policy (Jung et al., 2014).

More severe pollution problems emerged in the 1970s when the government promoted the export-driven heavy chemical industry. As global environmental movements started in the 1970s, some researchers and journalists started to discuss pollution. In 1977, the

Environmental Protection Act was established after President Park Chung Hee mentioned environmental issues in his new year's speech (Ministry of Environment, 2010). Jung et al. (2014) state that this was the first time that the government acknowledged that environmental pollution exists in the country. Environmental problems came to be regarded as a possible threat to economic growth and eventually to the elites who ran the developmental state institutions (Joo, 1999; Ministry of Environment, 2010).

Environmental impact assessment became mandatory for public development projects in 1981, but government officers often neglected the assessment process (Moon, 1997). The MOLIT did not conduct environmental impact assessments on many multi-purpose dam construction projects, and those projects were never stopped due to environmental reasons (Cho, 2006, p. 16). Environmentalists argued that although environmental impact assessments had become mandatory, they still indulged development (Jung et al., 2014 p. 291). Moreover, when environmental NGOs file a lawsuit against land development projects which do not follow legal process for environmental protection, they often lose the suit. Lee Younggi, a lawyer who has filed a number of environmental lawsuits, said that "environmental lawsuits are 100% a losing game (in Korea)" (Lee, Y., personal communication, 6 March 2017). The "judiciary branch poses few significant constraints" in the East Asian context under the developmental state paradigm (Moon & Prasad, 1994, p. 362).

4.5.2 Strong Presidency (Informal Institution)

A distinctive feature of Korean politics is that power is concentrated at the national level and the president has significant power over the policy-making process, including environmental policy. Together with the developmental state institutions, this strong presidency has played a critical role in water management policy.

Since the Environmental Protection Act was established in 1977, decisions on environmental regulations have been introduced by the national government. Once the national government enacts regulations, it is the responsibility of provincial and local governments to implement them with limited rights to modify (Jung et al., 2014). The sub-national governments are also tasked with the monitoring of environmental regulations. Many sub-national governments, however, tend to put a higher value on boosting the local economy than regulating environmental problems caused by local industries (Jung et al., 2014).

Although there were attempts to decentralise water management, the overall structure is still centralised. The structure of river basin management committees for the country's major rivers (the Han, Nakdong, Geum, and Yeongsan rivers) exemplifies the centralised governance. Established between 1999 and 2002, the river-basin committees deal with pollutant reduction plans, set water fee rates, and fund operations and land purchases in the basin areas (Jung et al., 2014, p. 179). Nevertheless, the chair of each river-basin committee is the Vice-Minister of Environment and the committees are mostly comprised of high-level officers from national-level institutions: the MOLIT, the Forestry Agency, K-water, the Korea Rural Community Corporation, the Korea Hydro, and Nuclear Power. This means that the main actors in the committees are still from the national level. The consultative committees within the river-basin committees are comprised of residents, NGOs, experts, and relevant companies, but these consultative committees cannot vote in the decision-making process.

In the case of the Four Rivers Project, most of the Korean interviewees said that the project started as a result of former president Lee's dream to build a Grand Canal and the developmental state institution's support for the president's plan. The task force on the Four Rivers Project was organised within the MOLIT. To support the task force, a committee of administrative support was established within the Office of the Prime Minister's Secretariat. In addition, local committees of the Four Rivers Project were established for each river basin.

> Presidents have been very powerful in South Korea. The power of presidents is even described as 'imperial presidency'. The President of South Korea can easily change enforcement ordinance. (Hong, J. H., personal communication, 21 March 2017)
>
> The power is concentrated in the hands of the president. The president can do whatever he/she wants. (Kim C. W., personal communication, 14 March 2017)

The strong presidency is connected to the weak participatory process. For example, the Low Carbon Green Growth Act, which provided legitimacy to the Four Rivers Project, did not include regulations regarding participatory governance in the decision-making process. Yun and Won (2012) argue that the green growth concept began to be discussed when President Lee suggested it. The agenda-setting process was led by the president and politicians who were seeking to recover their popularity after a series of protests against the government. Choi (2013) describes the communication process during the Four Rivers Project as a "monologue". While the government tried to hold events for advertising and organised catch-phrase contests, public hearing events were held at short notice (usually three days) and participants could not receive relevant information beforehand (Choi, 2013).

4.5.3 Divided Water Management Institutions (Formal Institution)

Water policy in South Korea has been largely divided into water quality and quantity policies and water management institutions have solidified such a system for a long time. The Ministry of Environment has taken care of water quality management while the MOLIT has been in charge of water quantity management. As a sub-agency of the MOLIT, K-water has driven water resource development, including dam construction projects. At the same time, environmental indicators of the Ministry of Environment do not include landscape conservation or the waterbody morphology rate (Ministry of Environment, 2019). In river management, the MOLIT is in charge of maintenance while the Ministry of Environment takes care of purification works. This divided structure was established in the early 1990s when a separate institution for environmental protection was created (Kim et al., 2017).

The Ministry of the Interior and Safety and sub-national governments also share tasks related to water management. Small river management is the work of the Ministry of Interior and Safety; at times their work overlaps with the MOLIT's responsibilities (Kim, 2011). Table 4.2

Table 4.2 Divided water management institutions (before 2018)

	Quantity (floods, water supply, water resource)	Quality (ecosystem conservation)
Ministry of Environment		The Basic Plan for Water Environment Management for the period between 2006 and 2015 Special Acts for the Nakdong River, Geum River, and Yeongsan River
Ministry of Land, Infrastructure and Transport (MOLIT), K-water (sub-agency of MOLIT) Ministry of the Interior and Safety	Long-Term Master Plan for Water Resource Long-Term Plan for Dam Construction (2012–2021) Integrated River Basin Flood Control Plan Small River Maintenance Master Plan Making Small Streams Beautiful and Safe	

summarises how water management policies are divided by ministry and agency. The policies were described separately in Sects. 4.3.1 and 4.3.2.

This segregated system has long been criticised for its inefficiency and there have been attempts to integrate the divided institutions (Kim, 2013, 2018a, b; Kim et al., 2017; Ahn, 2012). For example, a committee for water management policy adjustment was established in the Office of the Prime Minister's Secretariat in 1997 in order to coordinate water policies. The committee's decisions, however, did not carry a legally binding force, and the water management systems were not integrated. Another attempt was made in 2005 when the government decided to unify water management policies, but this plan was abandoned when the ruling party at that time (Democratic Party of Korea) was defeated in the 2006 local elections (Kim, 2011, 2018a, b).

The failure of water management integration can be attributed to the conflicts between different ministries that tried to secure and extend their own tasks. The attempts to transfer all water management tasks to the

Ministry of Environment faced strong opposition from the MOLIT (Kim et al., 2017). While the Ministry of Environment argued that it is necessary to integrate all water management tasks under its control, the MOLIT insisted that water management is part of land management. The MOLIT argued that they could not hand over dam construction and water quantity management tasks to other ministries (Kim, 2013). The Four Rivers Project exemplifies the impact of these divided institutions. The MOLIT drove the project planning and implementation while the Ministry of Environment was hardly engaged in the process.

Ironically, the impetus for water management integration was provided by the Four Rivers Project after its completion, as described in Sect. 4.3.3. The weirs constructed through the project have been heavily criticised for deteriorating the water quality of the rivers (Bae, 2019). After the ruling party changed from the Liberty Korea Party to the Democratic Party of Korea in 2017, one of the first directives announced by President Moon Jae-in was the integration of water quality and quantity management (Kim et al., 2017). Although the conflict between the Ministry of Environment, the MOLIT, and K-water intensified, integration was partially achieved. In 2018, most of the tasks related to water quantity management were transferred from the MOLIT to the Ministry of Environment, except the river management department ("Summary", 2018).

The three institutional features of the Korean water management policy investigated in this section, namely the developmental state, the strong presidency, and divided water management institutions, are in a way interconnected. The developmental state institutions were established during Park Chung Hee's military regime when the presidency was most powerful. By achieving fast-economic growth without investing in environmental protection, his regime could earn legitimacy and consolidate the strong presidency going forward. Even after democratisation in 1987, those institutional legacies remained, affecting environmental policy. In addition, as part of the developmental state institutions, the MOLIT and K-water could maintain their water quantity management task even though the Ministry of Environment was established in 1994. The institutional settings of the Korean water management policy analysed in this chapter are compared to that of Germany in the cross-case analysis in Chap. 6.

References

Ahn, K. S. (2012). A Study on Improvement of Overlapping Functions of the National Water Management Policy. *Korean Society and Public Administration, 23*(3), 183–208.

Ahn, H. (2014, December 23). Report of the Four Rivers Project Investigation Committee. *Joongang Ilbo*. Retrieved from https://news.joins.com/article/16764546

Bae, M. (2019, June 6). 10 years After the Four Rivers Project. *Kyunghyang Shinmun*. Retrieved from http://news.khan.co.kr/kh_news/khan_art_view.html?art_id=201906061726001

Board of Audit and Inspection. (2018). *An analysis of the Implementation and Outcome of the Four Rivers Project*. Board of Audit and Inspection. Retrieved December 26, 2019, from http://www.bai.go.kr/bai/groupreport/auditResults.do?mdex=bai222&rpstOrgCd=0000000024&audYr2=2017&audNo=509&dspRqSno=10&isSearch=1

Brand, K. (2015). Sustainable Development. In J. D. Wright (Ed.), *International Encyclopedia of the Social & Behavioural Sciences* (2nd ed., pp. 812–816). Elsevier.

Brand, U. (2012). Green Economy – the Next Oxymoron? No Lessons Learned from Failures of Implementing Sustainable Development. *GAIA, 21*(1), 28–32.

Byun, D.-G. (2011). A Review on the Four Rivers Restoration Project from the Viewpoint of the Collaborative Governance Systems for the Sustainable Development in Korea. *Journal of Social Science, 23*(2), 31–63.

Caldentey, E. (2008). The Concept and Evolution of the Developmental State. *International Journal of Political Economy, 37*(3), 27–53.

Chang, Y. (1996, February 28). Thirst After Five Years…Dams Are Urgently Needed. *Kyunghyang Shinmun*.

Chang, H., Franczyk, J., Im, E.-S., Kwon, W.-T., Bae, D.-H., & Jung, I.-W. (2007). Vulnerability of Korean Water Resources to Climate Change and Population Growth. *Water Science and Technology: Journal of the International Association on Water Pollution Research, 56*(4), 57–62. https://doi.org/10.2166/wst.2007.536

Cho, M.-R. (2003). Causes and Structure of Environmental Conflicts Concerning National Development Projects. *ECO, 2003*, 110–146.
Cho, M.-R. (2006). *Developmental Politics and Green Progress.* Environment and Life.
Choi, Y. (1970, February 14). Korea in 2000 (7) Water Resource. *Dong-a Ilbo.*
Choi, K. (1985, October 17). 3.4 Billion Ton Per Year of Water Can Be Supplied Now. *Dong-a Ilbo.*
Choi, C. (1990, December 25). The Dam Town, Gangwon, Rejects Dams. *Dong-a Ilbo.*
Choi, Y. (1995, September 23). Let's Build a Tunnel Which Connects Multi-Purpose Dams. *Dong-a Ilbo*, p. 23.
Choi, H.-K. (2013). Designing a Participatory Model of Environmental Policy Based on Scientific Technological Citizenship: Learning from the Swedish and Danish Cases. *Space and Environment, 23*(3), 6–51.
Climate Change Policy Research Institute. (2012). *Report on National Climate Change Policy in 2012.* Climate Change Policy Research Institute.
Ding, X. L. (1994). Institutional Amphibiousness and the Transition from Communism: The Case of China. *British Journal of Political Science, 24*(3), 293–318.
Environmental Conservation. (1981, November 2). Association Calls for Environmental Impact Assessment Before the Estuary Dam in Nakdong. *Dong-a Ilbo.*
Ferguson, P. (2015). The Green Economy Agenda: Business as Usual or Transformational Discourse? *Environmental Politics, 24*(1), 17–37.
Gladwin, T., Kennelly, J., & Krause, T. (1995). Shifting Paradigms for Sustainable Development: Implications for Management Theory and Research. *The Academy of Management Review, 20*(4), 874–907.
Government of Korea. (2009). *Green New Deal Strategy for Job Creation.* Government of Korea.
Government of Korea. (2010). *National Climate Change Adaptation Plan (2011-2015).* Government of Korea.
Government of Korea. (2011). *2011 Report on Extreme Weather Events.* Korea Meteorological Administration.
Government of Korea. (2015). *The 2nd National Climate Change Adaptation Plan (2016-2020).* Government of Korea.
Gwon, S. (1998, August 10). I Am Very Sorry…We Need a Fundamental Solution. *Kyunghyang Shinmun.*

Haggard, S., & Kang, M.-K. (2015). The Politics of Growth in South Korea: Miracle, Crisis, and the New Market Economy. In C. Lancaster & N. van de Walle (Eds.), *The Oxford Handbook of the Politics of Development*. Oxford University Press.

Han River Flood Control Office. (2010). *The Characteristics of Floods in Korea*. Retrieved March 1, 2020, from http://www.hrfco.go.kr/web/floodPage/floodClass.do

Hong, S. (2009, June 2). The Four Rivers Project Is River Development Without Restoration. *Hangyeoreh*, Retrieved April 23, 2020, from http://www.hani.co.kr/arti/society/environment/358140.html

Huber, J. (2000). Towards Industrial Ecology: Sustainable Development as a Concept of Ecological Modernization. *Journal of Environmental Policy & Planning, 2*(4), 269–285. https://doi.org/10.1080/714038561

Hwang, J. (2006a, October 25). Lee Myung-bak Visits Canals in Germany "I Will Visit the Daedong River Next Year." *Hangyeoreh*. Retrieved 1 January, 2021, from https://www.hani.co.kr/arti/politics/politics_general/167138.html

Hwang, S. (2006b). *Protest Against the Donggang Dam*. Retrieved April 1, 2020, from http://www.archives.go.kr/next /search/listSubjectDescription.do?id=003939&pageFlag=&sitePage=1-2-1

In Hyunwoo. (2021, April 14). Contrasting Opinions on the Results of Opening of the Dam Gates…Who Is Right? *Hankookilbo*. Retrieved 3 January, 2022, from https://www.hankookilbo.com/News/Read/A2021041415140004051

International Commission on Large Dams. (2019). *Number of Dams by Country Members*. Retrieved January 3, 2020, from https://www.icold-cigb.org/article/GB/world_register/general_synthesis/number-of-dams-by-country-members

Is the Onsan Illness False. (1985, March 14). *Joongang Ilbo*. Retrieved February 4, 2020, from https://news.joins.com/article/1983927

Jänicke, M. (1985). *Preventive Environmental Policy as Ecological Modernisation and Structural Policy*. International Institute for Environment and Society.

Jeon, Y. (2011). Analysis of the Residents' Understanding About Major Developmental Project. *Journal of the Korean Association for Policy Analysis and Evaluation, 21*(4), 167–198.

Johnson, C. (1982). *MITI and the Japanese Miracle*. Stanford University Press.

Johnson, C. (1987). Political Institutions and Economic Performance: The Government-Business Relationship in Japan, South Korea, and Taiwan. In F. C. Deyo (Ed.), *The Political Economy of East Asian Industrialism* (pp. 136–164). Cornell University Press.

Johnson, C. (1999). The Developmental State: Odyssey of a Concept. In M. Woo-Cumings (Ed.), *The Developmental State*. Cornell University Press.

Jones, R. S., & Yoo, B. (2011). *Korea's Green Growth Strategy: Mitigating Climate Change and Developing New Growth Engines*. OECD. https://doi.org/10.1787/5kmbhk4gh1ns-en

Joo, J.-H. (1999). Explaining Environmental Policy Development in South Korea: The Case of the Environmental Preservation Law. In *Korean Public Administration Review, 33*(1), 295–310.

Jung, C. (1972, November 25). *Korea's Largest Dam, Soyang Dam Completion Ceremony*. Dong-a Ilbo.

Jung, H., Lee, K., Jung, H., Kim, T., Chu, J., & Jeon, D. (Eds.). (2014). *Korean Environmental Policy*. Environment and Civilization.

Kang, J. (1976, July 6). Preventing Damage from Floods. *Kyunghyang Shinmun*, p. 7.

Kang, J. (2012). *The History of Dam Building in Korea: Korea National Committee on Large Dams*. Retrieved February 4, 2020, from http://www.kncold.or.kr/ds_imgs/sub04/40/266.pdf

Kang, M. G., Jeong, H. S., Lee, J. H., & Kang, B. S. (2013). Assessing National Flood Management Using a Sustainable Flood Management Framework. *Water Policy, 15*(3), 418–434. https://doi.org/10.2166/wp.2013.049

KFEM. (2000). *Onsan Disease*. Retrieved January 8, 2020, from http://kfem.or.kr/?p=38404

KFEM. (2009). *Ten Truths About the Four Rivers Project*. Retrieved April 23, 2020, from http://kfem.or.kr/?p=2357

Kim, T. (1970, October 16). Waterway from West Sea to Youngwol. *Dong-a Ilbo*.

Kim, M. (1986, September 8). New Han River and New Landscape. *Kyunghyang Shinmun*.

Kim, D. (2001). *Speech at Choong-chung College Graduation Ceremony. Presidential Archives*. Retrieved April 23, 2020, from http://www.pa.go.kr/research/contents/speech/index.jsp?spMode=view&artid=1309038&catid=c_pa02062

Kim, J. (2005). Long-Term Accumulation of Korean Economy and Economic Crisis. In K. P. A. Council (Ed.), *Change and Transformation of Korean Society: Historical Trajectory, Future in the Present and Academic Reproduction* (pp. 23–58). Hanwool.

Kim, Y.-T. (2008). *Bureaucrats and Entrepreneurs. The State and the Chaebol in Korea*. Jimoondang.

Kim, C. S. (2010a). The Dilemma of Development and Conservation. *Korean Society and Public Administration, 20*(4), 133–162.

Kim, C. S. (2010b). An Evaluation on 4 Major River Restoration Project: The Probabilities of Excluding Policy Errors. *Korean Public Administration Quarterly, 22*(4), 1041–1066.
Kim, H.-R. (2011). Review of Water Management Laws and Recommendations. *Public Land Law Review, 52.*
Kim, H. (2012, August 8). The Four Rivers Became "Green Tea Latte Disaster". *Nocut News*. Retrieved from https://news.naver.com/main/read.nhn?mode=LSD&mid=sec&sid1=100&oid=079&aid=0002383498
Kim, C.-H. (2013). Why Doesn't the Korean Water Management System Change? Based on Mahoney's Path Dependence Theory. *Korean Journal of Public Administration, 51*(4), 167–194.
Kim, J. (2017). *Chung, Doo-un "It's a Mistake to Inspect the Four Rivers Project". Interview with Doo-un Chung*. Retrieved November 10, 2019, from http://www.tbs.seoul.kr/news/bunya.do?method=daum_html2&typ_800=12&seq_800=10220014
Kim, J.-K. (2018a). *Integrated Water Management Policy for Future Generation.* Retrieved March 19, 2019, from http://www.waterjournal.co.kr/news/articleView.html?idxno=39650
Kim, J. (2018b, September 18). No More National Level Dam Project. *Hangyeoreh*. http://www.hani.co.kr/arti/society/environment/862560.html#csidx46580c8291df9b2869a997209e724a6
Kim, J. & Kim, Y. (2006, July 19). Multi-Purpose Dams Discussed After Floods Again. *Kyunghyang Shinmun*. Retrieved January 7, 2020, from https://news.naver.com/main/read.nhn?mode=LSD&mid=sec&sid1=102&oid=032&aid=0000185122
Kim, Y.-K., Yoo, J.-A., & Chung, E.-S. (2012). Water Management Vulnerability Assessment Considering Climate Change in Korea. *Climate Change Research, 3*(1), 1–12.
Kim, C.-K., Shin, H., & Kim, H.-J. (2015). Changing Environmental Attitudes and Neo-Developmentalism in South Korea. *Korean Society, 16*(1).
Kim, I., Kang, H., Kim, H., Kim, H., Moon, H., Ryoo, J., et al. (2017). *A Study on the Basic Principles and Policy Roadmap of Integrated Water Management.* Korea Environment Institute.
Korea Gallup. (2008a). *Evaluation of Lee Myung-bak Government's Performance.* Retrieved January 3, 2020, from https://panel.gallup.co.kr/Gate/Panel/F025.aspx?seq=15117&SearchGb=&SearchKey=&MSeq=15143&Yy=2008&PageID=F058&PGNO=1&date=Mon%20Feb%2025%2015:11:30%20UTC+0100%202019
Korea Gallup. (2008b). *Evaluation of Lee Myung-bak Government's Performance.* Retrieved January 3, 2020, from https://panel.gallup.co.kr/Gate/Panel/F025.

aspx?seq=15092&SearchGb=&SearchKey=&MSeq=15097&Yy=2008&PageID=F058&PGNO=1&date=Mon%20Feb%2025%2015:13:51%20UTC%200100%202019

Korea Meteorological Administration. (2014). *Korean Climate Change Assessment Report 2014*. KMA.

Korea Meteorological Administration. (2017). *Climate Change Projection Report in the Korean Peninsula for the New Climate Regime*. KMA.

Ku, D.-W. (1999). The Environmental Consciousness of the Korean People. *Journal of Environmental Policy and Administration, 7*(2), 17–33.

Ku, D.-W. (2011). Environmental Movement Case Study Through the Lens of Ecological Democracy. *Memory & Vision, 25*, 8–33.

K-water. (2015). *Water Encyclopedia*. Retrieved April 18, 2020, from http://www.water.or.kr/disaster/general/flood/flood03_datail06.do

Lee, S. (1979, August 15). Flood Control Measures that Changed the Map. *Kyunghyang Shinmun*.

Lee, K. (1984, September 22). I Will Be Targeted If I Stand Against. *Dong-a Ilbo*.

Lee, K. (1999, August 3). Multi-purpose Dam Planned for Imjin River. *Kyunghyang Shinmun*.

Lee, M.-H. (2005). Developmentalism and Korea Water Resources Corporations. *Democratic Society and Policy, 7*, 88–113.

Lee, C. (2006). *Tap Water Pollution Crisis*. Retrieved January 20, 2020, from http://www.archives.go.kr/next/search/listSubjectDescription.do?id=003912&pageFlag=

Lee, S.-H. (2010). Evaluation Over the Legitimacy of 4 Major River Project of Korea with the Viewpoint of Principle of Integrated Water Management. *ECO, 14*(1), 63–96.

Lee, S.-Y. (2011). A Comparative Analysis of the Koreans' Perception of Environmental Conflict: Based on the 2007 and 2010 Survey Results of Korean Conflicts. *ECO, 15*(2), 81–109.

Lee, J. (2012, August 7). Paldang Residents "No Need for Subsidies. We Want Deregulation". *YPDN*, Retrieved April 18, 2020, from http://www.ypdn.co.kr/paper/news/print.php?newsno=2951

Lee, J. (2014, March 27). 14 Years After the Cancellation of Youngwol Dam. *Hangyeoreh*, Retrieved April 7, 2019, from http://www.hani.co.kr/arti/society/society_general/630164.html

Mainwaring, S. (1992). Comparing Development in Brazil and Argentina - Kathryn Sikkink: Ideas and Institutions: Developmentalism in Brazil and Argentina (Ithaca and London: Cornell University Press, 1991. pp. xviii,

263. $35 00.). *Review of Politics, 54*(2), 333–335. https://doi.org/10.1017/S0034670500017952

Ministry of Culture, Sports and Tourism. (2008). *Green Growth, Let's Open the Age of Mature Freedom*. Retrieved December 26, 2019, from http://www.korea.kr/special/policyFocusView.do?newsId=148656226&pkgId=49500377&pkgSubId=&pageIndex=3

Ministry of Culture, Sports and Tourism. (2009a). *Final Confirmation of the Four Rivers Project*. Retrieved December 27, 2019, from https://www.mcst.go.kr/kor/s_notice/press/pressView.jsp?pSeq=9998

Ministry of Culture, Sports and Tourism. (2009b). *Awakening Four Rivers, Korea's Veins Are Being Restored*. Retrieved February 27, 2020, from https://news.naver.com/main/read.nhn?mode=LSD&mid=sec&sid1=117&oid=298&aid =0000018830

Ministry of Environment. (2008). *Final Report of Research on National Climate Change Adaptation Master Plan Establishment* (In Korean: 국가 기후변화 적응 마스터플랜 수립 연구). Ministry of Environment.

Ministry of Environment. (2010). *Three Decades of Environmental Policies in Korea*. Ministry of Environment.

Ministry of Environment. (2013). *List of Environmental Non-Governmental Organizations, Non-profit Organization, and Social Cooperatives*. Retrieved January 18, 2020, from http://www.me.go.kr/home/web/policy_data/read.do;jsessionid=tbY4yQ82bUHORYYzAWzD0ts+.mehome1?pagerOffset=1100&maxPageItems=10&maxIndexPages=10&searchKey=&searchValue=&menuId=10260&orgCd=&condition.deleteYn=N&seq=5076

Ministry of Environment. (2019). *2018 Annual Report of Environmental Statistics*. Retrieved February 1, 2020, from http://me.go.kr/home/web/index.do?menuId=123

Ministry of Environment. (2020). *Four Rivers Restoration Assessment Task Force*. Retrieved March 30, 2020, http://www.me.go.kr/rivers4nature/

Ministry of Land, Infrastructure and Transport (MOLIT). (2012a). *The Long-Term Plan for Dam Construction (2012-2021)*. MOLIT.

Ministry of Land, Infrastructure and Transport (MOLIT). (2012b). *The Integrated River Basin Flood Control Plan*. Ministry of Land, Infrastructure and Transport.

Ministry of Land, Transport and Maritime Affairs. (2009). *Masterplan for Four Rivers*. Ministry of Land, Transport and Maritime Affairs.

Ministry of the Interior and Safety. (2017). *2017 Contest for Making Small Streams Beautiful and Safe*. Retrieved January 2, 2020, from https://www.gov.kr/portal/ntnadmNews/1054019

Moon, T. (1997). *Environmental Policy Theories*. Hyungsul.
Moon, C., & Prasad, R. (1994). Beyond the Developmental State: Networks, Politics, and Institutions. *Governance: An International Journal of Policy and Administration, 7*(4), 360–386.
Natural Disasters and Man-Made Disasters Can Be Prevented. (1962, August 30). In *Dong-a Ilbo*.
Newsis. (2009). *Full Text of President Lee's Speech on Budget for 2010 at National Assembly*. Retrieved January 29, 2020, from http://news.chosun.com/site/data/html_dir/2009/11/02/2009110200586.html
NOAA National Centers for Environmental Information. (2019). *Climate at a Glance: Global Mapping*. Retrieved January 28, 2020, from https://www.ncdc.noaa.gov/cag/
OECD. (2018). *OECD Economic Surveys: Korea*. OECD. Retrieved November 11, 2019, from https://www.oecd.org/eco/surveys/Korea-2018-OECD-economic-survey-overview.pdf
Office of the Prime Minister's Secretariat. (2009). *Prime Minister Han "Four Rivers Project Is Not Grand Canal Project*. Ministry of Culture, Sports and Tourism. Retrieved March 1, 2020, from http://www.korea.kr/news/policyNewsView.do?newsId=148671927
Oh, C. (2019, March 17). Cloud Seeding and Mega-Purifier…Controversial Ultra-Fine Article Policy. *Hangyeoreh*. Retrieved from https://news.naver.com/main/read.nhn?mode=LSD&mid=sec&sid1=105&oid=028&aid=0002446719
Paldang Dam Completion Ceremony. (1974, May 25). In *Dong-a Ilbo*.
Park, C. (1968, July 23). Korean Landscape in 20 Years. *Dong-a Ilbo*.
Park, C.-K. (2009). Some Problems of the Green Growth Policy and the Project Saving the Four Major Rivers. *E&S, 83*, 117–360.
Park, S.-T. (2010). The 4-River Restoration Project from the Viewpoint of 21st Century River Management. *Korean Journal of Environmental Health Sciences, 36*(1), 72–75. https://doi.org/10.5668/JEHS.2010.36.1.072
Park, J. (2017, June 19). Gyeongin Canal that Costed 2.7 Trillion Won Is Empty. In *Nocut News*. Retrieved April 9, 2020, from https://www.nocutnews.co.kr/news/4800582
Pearce, D., Markandya, A., & Barbier, E. (1989). Blueprint for a Green Economy. *Earthscan*. https://doi.org/10.4324/9780203097298
Realmeter. (2010a). *Opposition Is Still Stronger Than Support for the Four Rivers Project*. Retrieved January 19, 2020, from http://www.realmeter.net

Realmeter. (2010b). *The Ruling Party's Failure: Four River Project, North Korea and Sejong-si*. Retrieved January 5, 2020, from http://www.realmeter.net

Sikkink, K. (1991). *Ideas and Institutions. Developmentalism in Brazil and Argentina*. Cornell University Press.

Song, Y. (2011, January 27). The Board of Audit and Inspection Did Not Inspect the Four Rivers Project Properly. In *Kyunghyang Shinmun*. Retrieved December 30, 2019, from https://news.naver.com/main/read.naver?mode=LSD&mid=sec&sid1=102&oid=032&aid=0002109093

Summary of the Integrated Water Management Policy. (2018). *Water Journal*. Retrieved February 1, 2020, from http://www.waterjournal.co.kr/news/article=View.html?idxno=41341

The Institute for Democracy. (2011). *Recommendation for Improving the Decision-Making Process of Major National Development Project*. The Institute for Democracy.

The Negative Impact of Development Projects Without Environmental Consideration. (1983, November 7). *Kyunghyang Shinmun*.

Woo-Cumings, M. (1999). *The Developmental State*. Cornell University Press.

Yang, J. (1993). Changes in Environmental Consciousness and Its Social Base. *Korean Journal of Sociology, 26*(4), 89–120.

Yang, S. (2011). An Analysis on the Interaction of Water Resource Policy Formative Process: Focused on the Grand Korean Canal Project of Lee, Myung-Bak Government. *The Korea Association for Policy Studies, 20*(3), 189–229.

Yoon, S.-W. (2016). The Historical Emergence and Reproduction of Growthism as Dominant Discourse in Korea. *Korean Society, 17*(1), 3–38.

Youn, S. J., Chung, H. B., & Choi, J. E. (2010). A Critical Study on "Green Growth" as Policy Directive for Environmental Law. *Environmental Law Review, 32*(2), 317–348.

Yun, S.-J. (2010). Not so Green: A Critique of South Korea's Growth Strategy. *Global Asia, 5*(2), 86–89.

Yun, S.-J., & Lee, D.-H. (2010). Agenda Setting and Frame of TV News About 4 Major Rivers Project in Korea. *ECO, 14*(1), 7–62.

Yun, S.-J., & Won, G.-Y. (2012). Social Acceptance of Lee Myung-bak's Green Growth-Based Climate Change Policy Regime: An Evaluation Based on a Survey of Experts' Perceptions. *ECO, 16*(2), 7–50.

5

Climate Change Adaptation and Water Management in Germany

5.1 Climate Change in Germany

5.1.1 Scientific Evidence of Climate Change in Germany

Germany has been experiencing rising temperatures and precipitation change due to climate change. Between 1881 and 2015, the average annual temperature rose by 1.4 °C (BMU/UBA, 2018). The number of frost days declined, and there has been a general increase in warm temperature extremes (Deutschländer & Mächel, 2017). A projection based on various emission scenarios and climate models suggests that the temperature will increase by 2.0 °C to 4.5 °C in the twenty-first century, compared with the period 1961–1990 (UBA, 2013). It is also expected that cold and warm extremes will occur more frequently in Germany (Deutschländer & Mächel, 2017).

Precipitation patterns have been changing. Located within the moderately humid climate zone, Germany has 860 mm of annual precipitation, although the rate varies by region (Hirschfeld et al., 2014). In the west,

Y. h. Kang, *Climate Change Adaptation in River Management*, Palgrave Studies in Water Governance: Policy and Practice,
https://doi.org/10.1007/978-3-031-10486-2_5

the average rainfall is between 650 mm and 1500 mm, whereas, in the east, it is between 450 and 650 mm (UBA, 2013). Climate change has aggravated the differences between the regions. The average annual rainfall in Germany increased by 11% between 1881 and 2016, but this change occurred primarily in the western part of Germany (UBA, 2013; BMU/UBA, 2018). The frequency of minor flooding has increased in the winter season in the southern river basins of Baden-Wuerttemberg and parts of Bavaria since the 1970s (UBA, 2013).

Precipitation will continue to change. Heavy rainfall is projected to become more frequent in Germany (BMU/UBA, 2018; Kunz et al., 2017). Annual precipitation is likely to increase by 9% until 2100 as compared to the average between 1971 and 2000. Notably, the frequency and intensity of heavy rainfall are expected to increase in winter. At the same time, summer is expected to become drier, with the dry period lasting longer and having more intensity. While precipitation in the winter months could increase by 17%, the summer months could have less rain (BMU/UBA, 2018). Notably, the eastern part of Germany has already experienced decreased rainfall in summer. According to the German Federal Environment Agency (*Umweltbundesamt*, UBA), this development is very likely to continue (UBA, 2013). Winters are likely to become wetter and warmer and summers hotter and drier (UBA, 2013).

5.1.2 Impact on Water Management

Climate change is creating various adaptation needs (Federal Government, 2008; Bauer & Steurer, 2015). As this book mainly focuses on flood risk, this section highlights the impacts of climate change on water management.

Precipitation and temperature change will affect river discharges. BMU/UBA (2018) states that less snow but more rain is projected in the mountain areas in winter. The spring melts will occur earlier than in the past as temperatures will increase. Such changes will lead to seasonal distribution changes in precipitation, affecting river discharges as a consequence. For example, the average discharge of the Rhine River in winter is increasing due to the increasing temperature, and this tendency will continue until the end of the century. At the same time, average and low water discharges in summer will decrease. As a result, the risk of both floods (in winter) and droughts (in summer) will increase in Germany (BMU/UBA, 2018).

The economic losses from floods have been more significant than any other natural hazards in recent years. According to the European Commission (EC), Germany had the largest total losses related to extreme weather events between 1980 and 2016 among EU member states (EC, 2018). For example, the floods along the Danube and Elbe rivers in August–September 2002 caused losses of 11.6 billion EUR, and the floods in the Elbe and Danube rivers in 2013 caused 6–8 billion EUR worth of losses (Thieken et al., 2016b; Marx et al., 2017).

Past experiences show that the magnitude of losses is not always proportional to the severity of floods. In Hamburg, for instance, the 1962 storm flood event caused significant damage, and 315 people were killed. Afterwards, flood protection measures and early warning systems were improved, and the damage was much less when another storm came in 1976 (Hirschfeld et al., 2014). Another example is Cologne, which was flooded in 1993 and damaged considerably. When even worse floods occurred in the same city two years later, the damage was less significant because the residents in Cologne were better prepared (Hirschfeld et al., 2014).

Besides economic losses, flooding has other considerable impacts on society, such as mental health and environmental damage. According to the European Environment Agency (EEA), coastal flooding in the EU could potentially cause five million additional cases of mild depression annually by the end of the twenty-first century under a high sea-level rise scenario in the absence of adaptation (EEA, 2016). Floods can also result in severe environmental consequences because facilities with toxic chemicals can be destroyed or inundated during flood events. Another possible damage to the environment due to floods is the destruction of natural habitats such as wetland areas (EC, 2016).

The low water level is another problem caused by climate change in Germany. The combination of higher temperatures and lower water levels can result in severe problems in the aquatic ecosystem. Water and soil temperature rise will lead to chemical and biological changes in water bodies (BMU/UBA, 2018). The solubility of oxygen is reduced at higher temperatures. In addition, reduced rainfall and increased evaporation during the summer season could dry out wetlands. The aquatic space for species would be reduced in a low water level situation. Furthermore, an extensive period of vegetation caused by climate change would increase the demand for water in agriculture. All of those changes may affect waterbodies significantly and disturb the aquatic ecosystem (BMU/UBA, 2018).

Alongside the impact on the ecosystem, the low water level can bring adverse impacts on navigation by reducing navigable days and transport capacity in summer. Germany has an extensive and well-developed inland waterway network. The length of the network is 7306 km, and about 90% of the waterways are used for shipping (UBA, 2013). When the water level is lowered due to climate change, the usability of the waterways can be reduced.

5.2 Adaptation Policy

5.2.1 German Strategy for Adaptation to Climate Change (DAS)

Climate change adaptation is now seen as a policy field of its own in Germany (Bubeck et al., 2016; Marx et al., 2017, p. 13). Adaptation was first not welcomed by policy actors in Germany because it was regarded as an 'inadequate' policy agenda which could hamper their strong commitment towards climate change mitigation (Hustedt, 2013). Nevertheless, the Federal Environmental Agency (UBA) started working on adaptation in 2002, stimulated by discussions in the UNFCCC and the EU, and research findings (Radtke et al., 2016; Stecker et al., 2012). Stecker et al. (2012) state that the UBA played an essential role in attracting the attention of environmental policy actors in Germany to adaptation issues.

Adaptation slowly became a more prominent policy agenda in Germany in the mid-2000s. The media discussed climate change impacts in relation to several extreme weather events including the 2003 European heatwave and floods in Austria in 2005 and 2006 (Stecker et al., 2012). The plan to establish an adaptation strategy was first announced as part of the Climate Protection Programme in 2005. This adaptation strategy was not in concrete shape, but it was the first step to formulate a national policy for adaptation.

The Federal Ministry for the Environment, Nature Conservation and Nuclear Safety (BMU; Bundesministerium für Umwelt, Naturschutz und nukleare Sicherheit) became more engaged in adaptation issues in late 2006 when the adaptation task was shifted from the division for international climate policy to the water policy division, which was closely linked to the EU water and adaptation initiatives (Stecker et al., 2012; Radtke et al., 2016). In the same year, the Competence Centre on Climate Impacts and Adaptation (KomPass) was established as a unit of UBA, following UBA's proposal to the BMU (Hustedt, 2013). The KomPass is tasked with policy advice, research, communication, and networking related to adaptation (Federal Government, 2008; UBA, 2019b).

The overarching framework for all the fields of adaptation is the German Strategy for Adaptation to Climate Change (*Deutsche Anpassungsstrategie*, DAS), which was adopted in 2008. DAS was prepared by a group of representatives from Federal Ministries and the group later became an inter-ministerial working group in charge of adaptation policy (IWG Adaptation Strategy, *IMA Anpassungsstrategie* in German), led by the BMU. The representatives of state governments (*Länder* in German[1]) were consulted over the course of the formulation.

The DAS follows a "standard approach" with comprehensive multi-sectoral adaptation strategies, which is preferred by OECD countries (Bauer & Steurer, 2015). Regarding water management, the DAS endorses the EU Water Framework Directive and the Floods Directive (Federal Government, 2008) (see more details in Sect. 5.3.1). The Adaptation Action Plan adopted three years after the DAS mainly deals with capacity-building activities (Bauer & Steurer, 2015). It is designed to be revised and updated following the results of regular evaluation (Federal Government, 2011).

[1] In this book, the German state governments are referred to as *Länder*.

The first evaluation report of the German Adaptation Strategy and the Action Plan undertaken in 2015 highlighted the advantage of natural water body structures. It was also pointed out that there is a considerable need for restoration of the rivers in Germany (Federal Government, 2015).

> Meandering rivers can slow down water flow and thereby reduce flood peaks. Also, riverbeds with sands and gravel can function as a buffer for the dry season because it allows exchange between surface water and groundwater. … Thus, the restoration of natural or near-natural waterbody structures through renaturation measures increases the adaptive capacity to climate change. … Many plants and animals are adapted to the specific habitat conditions of natural waters and a natural river dynamic and thus adapted to extreme conditions. As water temperature rises due to climate change, many aquatic organisms need to move to find suitable conditions. If the ecological passability (*Durchgängigkeit*) of watercourses is restored in the course of watercourse renaturation, these measures also support the adaptation to climate change.[2] (Federal Government, 2015, pp. 72–73)

This perspective is reflected in the adaptation indicators introduced by UBA in the same year. In regards to the water regime, one of the three response indicators is the "structure of water bodies" (Schönthaler & Andrian-Werburg, 2015, p. 30). The discourse related to water management found in the evaluation report will be analysed in Chap. 6 in comparison with the case of South Korea.

5.2.2 EU Adaptation Strategy

Adapting to climate change was put on the EU agenda in the mid-2000s. First, the *Green Paper on Adapting to Climate Change in Europe* was adopted by the EC in 2007 and the *White Paper—Adapting to Climate Change: Towards a European Framework for Action* in 2009. Although those two papers were not official legislation, they stimulated discussions on adaptation topics in the EU and influenced the adaptation strategies of the member states including Germany (Federal Government, 2008).

[2] The text was originally in German and translated for this book.

Built on such a gradual process of consultation and research, the EU Adaptation Strategy was adopted in 2013.

Although this adaptation strategy is not a legally binding policy document, its discourse on water management is worth looking into because the discussion in the EU has been influential in German environmental policy. It consists of eight actions regarding institutional settings for adaptation policy. The strategy especially encourages all member states to adopt "comprehensive adaptation strategies" (EC, 2013, p. 6). As losses and damages due to floods in the EU are expected to increase, the EC estimates that for every euro spent on flood protection, countries can save six euros from damage costs (EC, 2013). Green infrastructure and ecosystem-based approaches especially are encouraged in Action 7: Ensuring more resilient infrastructure. The EC states that it would endeavour to ensure the full mobilisation of ecosystem-based approaches to adaptation (EC, 2013). This is in line with the conservation-ecosystem service approach to water management which was introduced in Sect. 3.2.3.

The first evaluation report on the implementation of the EU Adaptation Strategy was published in 2018 (EC, 2018). Based on the opinions given by decision-makers and stakeholders at the national, regional, and local levels, the EC assessed that the strategy had put forward adaptation policies, particularly in cross-border areas within the EU such as river basins and Alpine areas (EC, 2018). Notably, the evaluation report emphasised the advantages of ecosystem-based adaptation. Taking the multiple benefits of this approach into account, the EC concluded that green infrastructure and nature-based solutions need to be further mainstreamed in investments and capital raising in the future (EC, 2018).

5.2.3 Adaptation in River Basins

Rivers are the primary component of the German landscape. There are ten river basins in Germany (Danube, Rhine, Maas, Ems, Weser, Elbe, Eider, Oder, Schlei/Trave, and Warnow/Peene) and the combined length stretches more than 400,000 km (BMU/UBA, 2018). This study concentrates on four of these rivers: the Danube, Rhine, Elbe, and Oder Rivers.

There are several reasons behind the selection. First, the four river basin districts account for the majority of German territory. Second, they are all international rivers shared with other countries. Third, the river basins experienced severe floods after the 1960s, and fourth, they make for a good comparison with the South Korean case, which deals with the Four Rivers Project.

This study focuses its analysis on the international and supra-regional commissions organised for river basin management in order to show the institutional feature of German river management. The commissions' discourse on flood risk management is also the subject of the analysis. International commissions have been organised for all the international rivers of Germany: The International Commission for the Protection of the Rhine (ICPR), the International Commission for the Protection of the Danube River (ICPDR), the International Commission for the Protection of the Elbe River (ICPER) and the International Commission for the Protection of the Odra (Oder) River against Pollution (ICPO). Furthermore, supra-regional commissions have been organised along the rivers such as the River Basin District (*Flussgebietsgemeinschaft*, FGG) Rhein, FGG Donau, FGG Elbe, and Coordinated River Basin District (*koordinierten Flussgebietseinheit*) Oder.[3] Adaptation policies are actively discussed in those international and supra-regional commissions beside the federal government and *Länder*. Below are the summaries of the adaptation policies adopted by the international commissions.

As extreme weather events are expected to occur more frequently in the Rhine River basin, the ICPR released the Strategy for the Rhine for adapting to climate change in 2015 (Krysanova et al., 2010; ICPR, 2015b). Regarding possible fields of action and measures, flexible win-win and no-regret measures such as renaturing, river bank strips, and securing floodplain areas free of use are preferred as adaptation measures (ICPR, 2015b). Concerning low flow management (drought-prone situation), the ICPR affirms that the measures should be compatible and coherent with flood risk management measures, such as floodplain reactivation and retention area construction.

[3] Because the river basin district commissions have their names only in German, the names of the rivers in the commission names are written in German in this dissertation.

In the case of the Danube River basin, the ICPDR (2013) projects that precipitation in the northern parts of the river basin is likely to increase while the southern parts would have less precipitation. Acknowledging such future predictions, the 'Danube Declaration' endorsed in 2010 by the ministers and representatives of the Danube countries indicated adaptation to climate change would be a focus area. The ICPDR adopted the first strategy on Adaptation to Climate Change in 2018 (ICPDR, 2019). This adaptation strategy is based on the Common Implementation Strategy Guidance Document No. 24 (River Basin Management in a changing climate) published by the EC (ICPDR, 2019). Produced in order to assist stakeholders in implementing the WFD, this guideline states clearly that eco-friendly, flexible, non-structural, and integrated measures need to be prioritised for river basin management (EC, 2009).

The ICPER and the ICPO have not set up specific plans for climate change adaptation, but their flood risk management plans recognise increasing flood risks due to climate change. The ICPER has established the "International Flood Risk Management Plan for the Elbe River Basin District" for 2016–2021. This plan follows the EU Floods Directive measures for flood risk management (ICPER, 2015). The ICPO has also set up the International Flood Risk Management Plan for the Oder River Basin District in accordance with the Floods Directive in 2015 (ICPO, 2015). Section 5.3.1 contains a more detailed explanation about the Floods Directive that influenced the river basin policies.

5.3 Water Policy

5.3.1 EU Water Framework Directive and Floods Directive

5.3.1.1 EU Water Policy Until 2000

Even before climate change adaptation became a policy agenda, useful policies for adaptation were established at various levels inside and outside of Germany. Particularly in the water management sector, relevant measures and policies were in practice without being classified as adaptation policies (BMU, 2017b; BMU/UBA, 2018). At the EU level, the European Water Framework Directive (WFD) and the Floods Directive are the most relevant policies for adaptation in water management. Although climate change was not the reason why those directives were established, they provide essential guidelines for adaptation. This section examines those directives in more detail, within the context of water policy development in Europe.

Water legislation at the European level started in 1975 when standards for surface waters (e.g., rivers and lakes) were set to regulate drinking water quality (Da-Cunha, 1989). Kaika (2003) describes those regulations as the 'first wave' of European water policy. They were based on command and control, and member states had little room to adjust the directives to their circumstances (Pahl-Wostl, 2015). Between 1988 and 1996, several directives, including the Urban Waste Water Directive, the new Drinking Water Quality Directive, the Nitrates Directive, and the Directive for Integrated Pollution and Prevention Control were adopted, making the period the 'second wave' of European water policy. This period came after the Sandoz chemical spill in 1986, and it was the first time that emission levels of chemicals started to be controlled by regulations (Kaika, 2003). In addition, the policies shifted from strict command and control to more adoption of voluntary instruments like the code of good agricultural practice. Nevertheless, problems were still dealt with in separate policies, and monitoring and reporting were not harmonised in this phase (Pahl-Wostl, 2015; UBA, 2013).

Subsequently, requests for more comprehensive and coordinated water legislation emerged from the European Parliament, member states, and public consultation (EC, 2019). As a result, European water management policy was brought to the current phase ('third wave') through the WFD which was adopted in 2000 (Directive 2000/60/EC) (EC, 2000). The WFD came into force in December 2000 as a legally binding policy, and it was transposed in the national legislation of member states. The WFD was implemented in 2012 when the programme of measures was made operational (EC, 2019). After the WFD, additional directives were adopted, but with the WFD as their basis.

5.3.1.2 Water Framework Directive

One of the significant changes introduced by the WFD to the EU is the concept of river basin management. This concept is an integrated approach to water management which covers from the origin spring to the estuary including tributaries. The WFD states that all member states are obliged to produce river basin management plans for each river basin district (Article 13).

Land use planning started to be connected with water resource planning through WFD (Kaika, 2003). In order to manage land use and water planning in an integrated manner, multi-level and intranational coordination is required among states and provinces along each river basin (Björklund et al., 2009). Besides, the WFD emphasises waterbody ecology and defines biological, chemical, and environmental objectives (Article 2). Member states are obliged to prepare plans and programmes to enhance the 'good status of water bodies' (BMU/UBA, 2018).

According to Pahl-Wostl (2015), the most notable point of the WFD is that it shows an evident change towards multi-level governance in water policy. It encompasses various instruments and yet allows more freedom to member states in implementation. It is also the first major directive adopted in the EU, which formally prescribes stakeholders' involvement (Pahl-Wostl, 2015). Stakeholder groups' participation was promoted as early as the development process of the WFD. Although the process was criticised for favouring organised groups (e.g., European Environmental Bureau, World Wild Fund, RSPB, and Birdlife), the open consultation made the decision-making process more transparent (Steyaert & Ollivier, 2007; Pahl-Wostl, 2015). Another criticism about the WFD is that the directive does not clearly indicate the participation process that is to be followed during the implementation phase (Kaika, 2004).

Twelve years after the WFD came into force, the EC published *Blueprint to Safeguard Europe's Water Resource* (2012), an evaluation report of EU water policy based on public consultation. It concluded that hydropower dams, navigation infrastructure, and land drains and embankments were still identified as the most widespread pressure on the water bodies (EC, 2012). Hydromorphological pressure (waterbody structure change) affects about 40% of the water bodies in the EU. Suggested alternatives to grey infrastructure include buffer strips and green infrastructure achieved through restoring riparian areas, wetlands, and floodplains which can reduce vulnerability to floods and droughts (EC, 2012). The EEA set a goal stating that during 2015–2050 a systematic approach to water management including climate-proofing, green infrastructure, and nature-based solutions should be followed (Künitzer, 2015).

5.3.1.3 Floods Directive

All the international river commissions in Germany (Rhine, Danube, Elbe, Oder, Moselle, Saar, and Maas) coordinate flood risk management in accordance with the Floods Directive (BMU, 2017b). The purpose of the EU Floods Directive, which came into force in 2007 (2007/60/EC), is to establish a framework for the assessment and management of flood risks. In this way, the directive aims to reduce the adverse consequences of flooding on human health, environment, cultural heritage, and economic activity (Article 7). The directive provides a three-stage approach to the member states: member states are required to (1) undertake a preliminary assessment of flood risks, (2) draw up flood hazard and risk maps, and (3) develop flood risk management plans based on the collected information (UBA, 2013). The assessment, flood hazard maps, flood risk maps, and flood risk management plans are subjected to review and update every six years.

It is notable that the Floods Directive suggests three phases for sustainable flood risk management: prevention, protection, and precaution (Article 7). Prevention measures are based on the precautionary principle. The methods include using natural floodplains by keeping these areas clear and removing or relocating existing structures from the floodplains (ICPER, 2015). Protection measures refer to maintaining or restoring natural water retention areas in order to slow down runoff. Finally, precaution measures include flood forecasting and warning, emergency measure planning, and increasing public awareness (BMU/UBA, 2018). In all three phases, the emphasis is on non-structural and green and blue infrastructure. This is in line with the tendency for synergising flood risk management and ecosystem protection in Europe since the 1990s (Kaczmarek et al., 1996) (see Sect. 3.2.2).

5.3.2 Federal Water Act

The German Federal Water Act (*Wasserhaushaltsgesetz*, WHG) was first legislated in 1957 and has been amended several times. The Seventh Amendment undertaken in 2002 was one of the most critical amendments because the EU WFD was transposed into the WHG. In 2009, the WHG was amended again for transposing the Floods Directive into German law (BMU/UBA, 2018). In fact, the federal government used this last amendment of the WHG as an opportunity to legislate more detailed policies at the national level after the 2006 federal reform (see Sect. 5.5.1). The updated WHG entered into force on 1 March 2010 (UBA, 2013).

The WHG specifies necessary provisions regarding water quality and quantity management. In general, water bodies including surface water, groundwater, and coastal water are subject to government control. Besides provisions for water quality control (wastewater treatment, water conservation, and water protection area designation), provisions for preventive flood mitigation measures are regulated by the WHG. The implementation of the WHG and other provisions relating to water are under the responsibility of *Länder*. The federal waterways are the exception as the federal government is responsible for the maintenance, development, and traffic requirements of these waterways (UBA, 2013).

Hydromorphology has gradually gained more attention in the federal regulations as a way to improve the ecological balance of water bodies (UBA, 2013). In impounded water bodies, there is a higher possibility of algal bloom, sludge accumulation, and oxygen deficiency. Particularly, the 7300 km-long canals built between the eighteenth and twentieth centuries contributed to the deterioration of water quality (UBA, 2013). The federal government used to prioritise water quality management in the past, but now it values integrated water management in order to minimise the impacts of river engineering measures (UBA, 2013). In the general principles of the current WHG, the precautionary principle and the integrated approach to ecosystem protection are emphasised. Besides, keeping natural conditions of water bodies is ensured as a flood prevention method (UBA, 2013).

5.3.3 The National Flood Protection Programme (*Das Nationale Hochwasserschutzprogramm*)

Floods are natural events, which are not uncommon in rivers, coasts, and flood plains. As land use changed for human societies, however, floods became a natural hazard. Many natural flood plains and wetlands have been turned into industrial, residential, or agricultural areas, and river courses have been straightened or shortened for navigation or hydropower. Such changes resulted in faster flow and increased volumes of water per unit during flooding (UBA, 2013). In Germany, heavy modification of rivers has been predominant in the last 250 years. For example, 87% of flood plains in the Upper Rhine were lost due to large-scale straightening since the mid-nineteenth century. Only about one-third of the former flood plains can function as retention areas in case of severe floods in Germany (UBA, 2013).

Although floods have occurred throughout the history of Germany, the recent floods in June 2013 particularly influenced the current flood policy in the country. Severe and widespread river floods affected 12 out of 16 *Länder*, and 8 *Länder* declared a state of emergency (Thieken et al., 2016b; Marx et al., 2017). In September 2013, three months after the floods, the development of the National Flood Protection Programme (NFPP) was decided at the conference of the environment ministers of the federal government and *Länder*. Other requests emerged during the conference, such as adjustment and modification of the legal framework regarding flood risk management, review of flood forecast, and more use of private insurance (Schwarz, 2015). The NFPP was then developed by the experts from the federal government and *Länder* in the Working Group of the Federal States on Water Issues (LAWA) and the National/

Federal Consortium for Nature Protection, Landscape Conservation and Regeneration (LANA) (BMU/UBA, 2018).

Adopted in 2014, the NFPP intends to strike a balance between the burdens and benefits experienced upper stream and downstream under the solidarity principle (BMU, 2017b). Germany's Basic Law states that flood protection is a *Länder*'s task and each *Länder* has its own policy (BMU/UBA, 2018; Thieken et al., 2016a). Some *Länder* already had their own climate change adaptation measures for flood control before the NFPP or the DAS. For example, Bavaria and Baden-Württemberg incorporated climate factors in their river dikes, and Schleswig-Holstein in their storm flood dikes (UBA, 2013). However, the floods in recent years strengthened the idea of solidarity among *Länder,* and this was the motivation behind the establishment of the NFPP (BMU, 2017b).

In the NFPP, measures are categorised into three: regulated retention measures, dike shifts, and elimination of dike weak points. In addition, effectiveness, synergies, feasibility, environmental friendliness, and costs are the criteria for measuring prioritisation (Schwarz, 2015). As of 2018, the NFPP scheme defined 32 cross-state projects consisting of 80 sub-projects for dike relocation and 59 sub-projects for managed flood retention (BMU/UBA, 2018). More than 1180 million m^3 of retention volume and more than 20,000 hectares of flood plains are expected to be created through the projects. The budget for the projects is approximately 5.5 billion EUR (BMU/UBA, 2018; BMU, 2017a).

Notably, the NFPP focuses on giving more space to the rivers, aiming for synergies between flood protection and nature conservation. To consider a river basin as a whole is the key concept behind the NFPP. "Preventive Flood Protection", a special framework for financing cross-state measures within the NFPP, intends to give more room to the river (BMU, 2017b). Such an approach is in line with the WFD and is also reflected in the National Adaptation Strategy (DAS). According to a five-year research project conducted by the Federal Institute of Hydrology (BfG) on the effectiveness of the NFPP, the NFPP measures could significantly reduce the water levels of future floods (UBA, 2021).

The role of the federal government in flood policy grew more prominent through the NFPP as the federal government can now coordinate the priority selection process. As supra-state flood control measures are the core of the NFPP, the role of the federal government as the coordinator has become more critical. Before the NFPP was established, there were no uniform assessment criteria for flood protection measures in Germany because the task belonged to the Länder. The NFPP is the first national policy that sets the priority for flood protection measures and provides a framework for cross-state flood protection (BMU, 2017a).

In November 2016, the federal government adopted the draft of the Flood Protection Act II in order to provide the legal basis for the NFPP. This act intends to simplify the procedures needed for implementing flood protection measures (BMU, 2017b). It is expected that the act will facilitate and accelerate procedures for the planning, authorisation, and construction of flood protection infrastructure, while underpinned by a participatory process (BMU, 2017b).

5.3.4 Critical Junctures

German water management policy has changed over time. In this section, the multi-level institutional changes in water management policy between 1950 and 2019 and the critical events that influenced such changes are reviewed from a historical institutional perspective. News articles from two major media in Germany, namely *Der Spiegel* and *Frankfurter Allgemeine Zeitung*, were used for finding critical events and policy changes over the period mentioned above.

Although disasters such as chemical accidents and extreme flood events are often referred to as windows of opportunity for evaluating the effectiveness of water management measures (Bubeck et al., 2015), events such as these with the potential to shock a system, do not always lead to policy change. When floods occurred in Bavaria in 1954 and Hamburg in 1962, grey infrastructure measures like dike strengthening were emphasised just like before the floods (Lücken im Deich, 1954; Deiche statt Harem, 1962). After a flood occurred in Saarbrücken in 1970, the Federal Minister of Transport promised that the Saar River would be canalised so that such a catastrophe would not happen again (Kanalisierung versprochen, 1970). In the case of water pollution, when transnational river pollution became politically prominent in the 1970s, Germany vetoed the proposal of the Netherlands which required pollution reduction in the Bathing Water Directive (Wurzel, 2002).

Discussions on alternative methods for floods first appeared in the 1970s. A Hamburg-based engineer Herbert Gudehus argued that higher dikes built in Hamburg after the 1962 floods resulted in more frequent storm surges in the area. His argument was first regarded as 'absurd' by many engineers (Irgendwo mutt dat Woter jo hen, 1976). After the 1973 flood in Hamburg, however, local companies sued the state government for damages citing Gudehus' paper as evidence. In 1976, Hamburg was flooded again, and more doubts arose around the effectiveness of dikes. When floods occurred in 1978 in Baden-Württemberg, the importance of floodplains emerged in the media (Ländle unter, 1978).

In the 1980s, the argument about the negative effect of river engineering gained more support within academia. Although some hydraulic engineers argued that large-scale flood catastrophes occurred without human influence, water economists and ecologists gradually concluded that the influence of urbanisation, agriculture, and industry on flood risks could not be denied (Dromedar im Kanal, 1983). They claimed that water flowed too fast due to canalisation, concreting of riverbanks, and straightening of streams (Bewerunge, 1988). In addition, environmental movements grew significantly in the 1980s and environmental NGOs promoted preventive flood protection and natural floodplains as an alternative. Bundes Naturschutz, an environmental NGO, argued that river floods were 'homemade' after floods occurred in Bavaria in 1980 (Hochwasser in den Flüssen zu einem Gutteil hausgemacht, 1980). The Auen Institute established by the WWF in 1985 was the representative organisation which supported changes in flood risk management policy (Knauer, 2004).

Besides severe floods, the Sandoz chemical spill in 1986 generated momentum for policy change in water management. The news of the Sandoz accident spread quickly and became one of the major environmental accidents in modern Europe. The fact that the chemical accident occurred only a few months after the Chernobyl disaster contributed to the rapid and sensational reaction towards it. Together with chemical pollution, river engineering methods, including dams and canalisation, were criticised for water quality degradation. Anne Schulte-Wülwer-Leidig, who was the Executive Secretary of the ICPR described during the interview that the reaction to the Sandoz accident was unprecedented:

> The change [in momentum] in Western Europe was really the Sandoz accident. There's another severe accident at the end of the 1960s. In 1969, toxic substances [were released] from near Frankfurt, and all the fish were killed in the downstream up to the Netherlands. There was news on TV but no reaction from the countries … in 1986, the Chernobyl disaster occurred and Sandoz on 1 November. The reaction was very fast among politics and the public at that time. A lot of demonstrations on the bridges [were organised] in different cities. There were three ministerial meetings within 11 months—the first ministerial meeting was held after one week of the accident in Zurich near the disaster place. Then the sanitation programme, Rhein Action Programme was adopted on 1 October 1987. It was a "wake-up call" at that time for politics, public, and industry. (A. Schulte-Wülwer-Leidig at ICPR, personal communication, 27 October 2017)

Reflection on hydraulic engineering and criticism against it entered into politics after the floods in the early 1990s. German President Roman Herzog stated that "[w]hat happened in hydraulic engineering in the past is proving to be incorrect today" after the 1995 floods (Alles harmlos, 1995). While some *Länder* began to reverse the previous river management policy by implementing restoration projects, changes at the federal level policy were slower. Chancellor Helmut Kohl expressed his opinion on the importance of 'giving more space to the rivers' after the 1997 Oder flood, but the policy did not change immediately. One of the main reasons behind the slow change was that regaining floodplains from agricultural or urbanised lands was not popular among residents and farmers (Bojanowski, 2013). The Auen Institute expected that "there will be high psychological and political hurdles to overcome" if the politicians try to change the flood policy from grey infrastructure measures to floodplain extension (Knauer, 1997).

Thieken et al. (2016b) state that the 2002 Danube and Elbe floods exemplify the role of critical events in the policy process. In 2002, 131 dikes failed to prevent the floods in Saxony, causing 21 fatalities and an economic loss amounting to 10 billion EUR (Thieken et al., 2016b). The unclear responsibilities and task distribution between different government levels were pointed out as the primary reason for the failure of flood protection (Thieken et al., 2016b). A former Minister of the BMU, Jürgen Trittin, emphasised the role of the federal government after the floods.

> Only the federal government can ensure uniform standards and only the federal government can create conditions for a fair balance between upper and lower riparian. The role of the federal government must, therefore, be strengthened. This is the only way to achieve a joint flood programme between the federal government, the Länder and the neighbouring countries—Jürgen Trittin in an interview with *Der Spiegel*. (Bölsche, 2002)

In 2005, the Act to Improve Preventive Flood Control (*Hochwasserschutzgesetz*) was introduced as the first nationwide requirement for flood damage prevention (BMU, 2005). While the task of flood control still remains at the *Länder* level, the regulation states that "if an agreement on a flood control measure cannot be reached in the context of the cooperation [among *Länder*], on request of one of the *Länder* the federal government shall intermediate between the *Länder* concerned" (Act to Improve Preventive Flood Control of 2005 Article 32). After the major federal reform in 2006, the federal government became able to legislate more detailed regulations for water management (UBA, 2013).

Nature-based solutions to flood risk management have been institutionalised since the 2000s. The Act to Improve Preventive Flood Control states that *Länder* shall adopt regulations to "protect or improve the ecological structures of bodies of water and their flood areas" and "preserve or claim and particularly recover natural retention areas" (Article 31b). After the 2013 Danube and Elbe floods, a dispute about flood risk management measures escalated as conservationists demanded stronger regulations for floodplain protection (Ehrenstein, 2013). The NFPP adopted in 2014 shows that policy change has occurred. As can be seen in Sect. 5.3.3, the NFPP emphasises 'giving more room to the rivers' for more effective flood management as well as synergies with nature conservation (BMU, 2017b).

Table 5.1 Critical events and institutional changes related to German water management between 1950 and 2019. Critical events are italicised and policy/institutional changes are underlined

	Federal	River Basin Commission	EU	International level
1950		ICPR established		
1957	Federal Water Act enacted			
1972	Constitution amendment			
1975			Bathing Water Directive	
1986	BMU created		*Sandoz chemical accident in the Rhine*	*Chernobyl disaster*
1987		Rhine Action Programme by ICPR		
1990	Re-unification	ICPER established		
1992				Rio Summit UNECE Water Convention adopted
1993	*Rhine floods*			
1995	*Rhine floods*			
1996		ICPO established		
1997	*Oder Flood*			UN Convention on the Law of the Non-Navigational Uses of International Watercourses adopted
1998		ICPDR established		
2000			*Water Framework Directive (WFD)*	

(*continued*)

Table 5.1 (continued)

	Federal	River Basin Commission	EU	International level
2002	*Danube and Elbe floods* Federal Water Act reform for WFD transposition			
2005	Flood Prevention Act			
2006	Federal Reform			
2007			*Floods Directive*	
2009	Federal Water Act reform for Floods Directive and more legislative power at federal level			
2013	*Danube and Elbe floods*			
2014	National Flood Protection Programme (NFPP)			

Table 5.1 is the summary of the critical events and institutional changes related to German water management policy. The policy changes through German water management history, however, have been not only the result of shocking events. Many interviewees pointed out the influence of social discourses when they were asked about the main cause of policy change. In the following section, the dominant discourses that have affected water management are reviewed and analysed.

5.4 Discourse Analysis

5.4.1 Optimism for Technology

Belief in science and technology has a long legacy in Germany. Technological advancement has connoted positive change in the country. Optimism for technology is not always on the opposite side of environmentalism. The advancement of technological solutions for pollution was one of the critical preconditions for the rise of environmentalism in Germany in the 1980s (Uekötter, 2014). Environmentalists and engineers could agree on promoting green technology. Ecological modernisation that emerged in Germany in the 1980s particularly embraces this idea (Jänicke, 1985; Huber, 2000). Pollution abatement technology provided a significant number of jobs and made it possible to export technology to the global market (Wurzel, 2002). Uekötter (2014) highlights that green technology was supported by politicians because "there was usually a notion in the air that this would boost markets for technology made in Germany" (p. 152).

David Blackbourn's book *The Conquest of Nature: Water, Landscape and the Making of Modern Germany* (Blackbourn, 2006) offers an insight into how strong belief in technology has significantly influenced river management in Germany. Until 200 years ago, rivers in Germany such as the Rhine looked very different from today. Floodplain, high peat moor, and wetland forests were mostly intact along the rivers. Sandbars, gravel banks, and islands divided the flow in numerous ways and flow speed varied by seasons (Blackbourn, 2006). The course of the Upper Rhine was particularly unpredictable. The flood triggered by spring snow melts or summer rains could sweep away villages (Hairsine, 2017).

From the mid-eighteenth century, however, Germans transformed their landscape rapidly. Lands were reclaimed for farmland and industrial facilities, while rivers were canalised for navigation as well as for flood control. Dams were built in hundreds of high valleys (Blackbourn, 2006; Federal Government, 2015). Johann Gottfried Tulla is one of the most famous engineers in Germany because he led the Rhine Correction programme which included straightening the river (Hairsine, 2017).

According to Blackbourn (2006), there had been previous attempts to change the river flow, but the scale and impact of hydrological projects from the eighteenth century were unprecedented. The river engineering in Germany was even more accelerated in the 1950s–1960s when the economic growth rate in West Germany was high, under the pressure of reconstruction after the Second World War (Blackbourn, 2006). The largest dams in Germany, such as the Rosshaupten Dam on the Lech River and the Bigge Dam in the Sauerland, were built in this period (Blackbourn, 2006). Furthermore, it was estimated that 40,000 kilometres of rivers and streams were straightened and diverted between 1945 and 1983 (Dromedar im Kanal, 1983). Blackbourn (2006) explains that many dam construction and river engineering projects were undertaken because of the 'utopian' discourse on technology.

> Read the utopian claims in 1900 about hydro-electric power, a clean and modern source of energy created by men in white coats, and it sounds just like the enthusiasm for nuclear power sixty years later. What did not change is the basic idea that nature was an adversary to be manacled, tamed, subjugated, conquered and so on through a dozen variations. (Blackbourn, 2006, p. 3)

The optimism for science and technology was universal regardless of political ideology in Germany. Liberals and socialists had a consensus when it came to hydrological engineering projects (Blackbourn, 2006). Such a view was shared among not only the elite but also the public. Large-scale civil engineering projects were welcomed with enthusiasm in most cases.

Optimism on hydrological engineering projects was firmly grounded in various outcomes (Blackbourn, 2006). The reservoirs created through dam construction, for example, played an essential role in drinking water supply, energy generation, flood prevention, and recreation space provision (BMU/UBA, 2018). Furthermore, new land gained from draining wetlands and moors was necessary to support a growing population. Malaria could be removed through such projects. Faster transportation of people and goods meant more freedom for the local communities. The transformation of landscape was therefore regarded as the conquest of nature and celebrated as progress (Blackbourn, 2006). Rivers were treated as 'machines' or 'resources' for economic growth and prosperity. Many interviewees described the prevalent discourse which influenced water management in Germany as below:

> There were many weirs in the Upper Rhein built at the end of the nineteenth century and the twentieth century. For hydropower generation, you have to change everything in the river. At that time, it was possible because only the economy was important. (A. Schulte-Wülwer-Keidig at ICPR, personal communication, 27 October 2017)
>
> In the early twentieth century, the whole river system in this area was changed to an artificial system. There were several reasons why they did it. The lowering of the surface was necessary due to [shipping for] coal mine industry. Also, the population increased a lot: from 0.5 to 3.5 million. (E. Pfeiffer at Emschergenossenschaft/Lippeverband (EGLV), personal communication, 7 November 2017)

Impounded rivers became part of the German landscape. Two-thirds of former floodplains have disappeared in the last 250 years (Brunotte & Amberge, 2009). Floodplain loss was more significant in large river basins including the Rhine, Elbe, Danube, and Oder: there are only 10–20% of former floodplains left on those rivers (BMU/UBA, 2018). The altered Rhine became the symbol of German identity (Blackbourn, 2006). Pahl-Wostl (2015, p. 15) points out that "idealised design principles based on technological panacea" have been applied to water management without consideration of long-term effects.

River engineering with grey infrastructure resulted in the degradation of ecological status in the rivers. Only 10% of waterbodies in Germany have good ecological status under the WFD criteria (BMU, 2010). Loss of natural habitats, algal bloom, sludge accumulation, and oxygen deficiency are common problems that can arise from waterbody impoundment (BMU/UBA, 2018).

The interviewees of this book mentioned that optimism for technical measures is still active in today's German society. According to them, while many water engineering experts now understand the limits of technical measures for flood protection, many policymakers, as well as the public, support, and trust such measures.

> There was a generation which thought you could handle all problems by technical approaches alone. But we technicians (now) understand that there are limits. It's fascinating that non-technicians like economists or bankers have this old-fashioned economic model until today that you can fix all problems by technology. And we, technicians who construct big dams, bridges and all the stuff, are sometimes [speechless (author added)]. Sometimes, people who know best know there are limits. (M. Grambow, Head of the Water Management and Soil Conservation Department in the Bavarian Ministry of the Environment and Consumer Protection, personal communication, 8 November 2017)
>
> The large area of Germany is all protected by dikes—like in the Netherlands. People trust these dikes. Sometimes the dikes are higher than the houses. (J. Hirschfeld, personal communication, 16 October 2017)

Beliefs in technology and environmentalism have sometimes been in conflict and other times in cooperation within German environmental politics. For example, the energy transition policy shows how the cooperation between environmentalism and optimism for technology could work. In the water management sector, however, engineering methods such as dams and dikes are usually on the opposite side of nature conservation. Flood risk management represents the battleground of the two different approaches.

5.4.2 Environmentalism

Although Germany is one of the pioneering countries in modern ecological ideas, development impetus was not stopped by ecological concerns until the 1960s. For example, the term 'ecology' was first coined by a German zoologist Ernst Haeckel in 1866 (Blackbourn, 2006). Nevertheless, when air pollution due to a toxic concentration of pollutants killed several people in Meuse Valley, Belgium in 1930, a German official in charge of air pollution control declared it as 'not a significant event' (Uekötter, 2014, p. 3). State-level air pollution laws existed in the

1960s, but the national-level environmental policy based on environmental protection was not developed until 1969 (Jänicke & Weidner, 1997).

During the intensive economic period after the Second World War, lack of environmental policy was a problem in water management. Major rivers were treated as natural sewers (Wurzel, 2002; Weale, 1993). The self-purification capacities of waterbodies could not function anymore by the end of the 1960s because of the excessive amount of artificial fertiliser, toxic chemicals, metal sediments, oil spill, and household wastewater. Fish caught in the major rivers were unsafe to eat. Blackbourn (2006) describes the situation as: "The fish-kills ended in Bavarian rivers only because there were no more fish to kill. The same problem eventually overtook the major rivers that had to carry this toxic load to the sea. (…) The Rhine was close to being biologically dead" (pp. 311–312). The industries that poured toxic substances were not called to account for water pollution. When the federal government set up an expert commission in 1959 after vast mountains of foam were floating in the summer, the detergent industry was included as a member of the commission in addition to scientists and water agencies (Uekötter, 2014).

Concerns over pollution and environmental protection grew in Europe in the 1960s. The Club of Rome was created in 1968 to examine global problems such as socio-economic development and environmental degradation. The Council of Europe adopted the Water Charter and designated 1970 as the European Nature Protection Year. In addition, the UN Conference on the Human Environment held in Stockholm in 1972 showed that environmentalism had become a significant discourse in international society (Jänicke & Weidner, 1997).

Such discursive transformation in Europe affected Germany. Economic consideration was still dominant in German politics, but environmental movements grew slowly. Several citizens' initiatives for environmental protection were organised particularly against nuclear power in the 1970s, building the basis of green politics (Dryzek et al., 2002; Rucht, 1990).

The new government coalition, which was formed by Free Democrats (FDP) and Social Democrats (SPD) after the 1969 election, influenced substantially the growing environmental discourse (Jänicke & Weidner,

1997; Blackbourn, 2006). Under the new government, new laws were passed for stronger environmental protection. Two opinion polls conducted in West Germany showed a dramatic change in public awareness. In September 1970, only 40% of the respondents said they were familiar with the term 'environmental protection'. One year later, the rate rose to 90% (Blackbourn, 2006, p. 317). The government established the Federal Environmental Agency (UBA) in 1974.

Uekötter (2014) describes the 1980s as the period when a boom of German environmentalism started. The term '*Waldsterben* (forest dieback)' became a popular term that alerted the public in the early 1980s. This word indicated a large-scale forest decline in Germany and other European countries. Industrialism which resulted in pollution including acid rain was blamed as the cause of the environmental destruction (Uekötter, 2014). Within the context of enhanced public awareness about the environment, the Ministry of the Environment, Nature Conservation, and Nuclear Safety (BMU) was established in 1986 less than two months after the Chernobyl disaster.

One of the most noticeable changes in the 1980s in German environmental politics is the establishment of the green party, *Die Grünen* (the Greens) in 1980. Although the party earned only 1.5% of the vote in the 1980 election, the share of the vote won by the party increased significantly during the 1980s. The party gained 8.3% of the vote and 44 seats in the 1987 federal election (Federal Returning Officer, n.d.). With success in elections, *Die Grünen* has played a vital role in environmental policy development. While the political parties, the government, and industry were quite hostile towards the environmental movement until the 1970s, major parties started to feel the pressure to adopt 'green positions' so as not to lose votes to *Die Grünen* (Dryzek, 2013; Uekötter, 2014; Blühdorn et al., 1995). Jänicke and Weidner (1997) point out that even the conservative media which had a close relationship with industry had changed their attitude to environmental issues by the late 1980s. Industry became hesitant to stand up against BMU, fearing a bad reputation (Wurzel, 2002). The impact of *Die Grünen* went beyond Germany, influencing green movements and political thought in other countries too (Schreurs & Papadakis, 2020; Blühdorn et al., 1995).

Germany is also the first country that adopted the precautionary principle (*Vorsorgeprinzip*) in environmental policies (Dryzek, 2013). This principle intends to prevent any dangers to the environment from arising in the first place. Scientific uncertainty should not be an excuse for inaction when it comes to environmental problems, according to the principle. The principle was stated in reports on environmental issues published by the federal government (West Germany) in 1976. The government proclaimed that precautionary environmental protection is a guiding principle of environmental policy (UBA, 2019a). Such policy can be compared to Britain and the United States where the national governments resisted adopting the principle (Dryzek, 2013).

In contrast, pollution was not regulated in the German Democratic Republic (East Germany) due to a lack of resources and political will. While the Constitution of East Germany had a clause on nature protection and pollution abatement plans, the law was not implemented in practice (Schreurs & Papadakis, 2020). After the reunification between East and West Germany in 1991, cleaning contaminated land as well as reducing water and air pollution in former East German states became the most urgent issue for the federal government (OECD, 2001).

Environmental protection became a goal in the German constitution by the mid-1990s (Dryzek, 2013). The Constitution included a commitment to protect the natural foundations of life for future generations (Schreurs & Papadakis, 2020). By the 2010s, the government of Germany showed stronger commitment and performance in climate protection than other industrialised countries (Dryzek, 2013).

With rising awareness of the environment, "the very projects that once symbolised human emancipation and progress" became "human and environmental catastrophes" (Blackbourn, 2006, p. 10). Hydrological projects such as dam and canal construction were seen as the cause of damage to ecosystems, loss of communities, and a threat to human health. New findings from hydrology contributed to the discursive change as well. Studies showed that regular local flooding occurred less when rivers were straightened, but once floods occurred, the results were far more devastating (Blackbourn, 2006). Many interviewees pointed out that the policy change in water management can be very much attributed to discursive change.

> [The policy change] was strongly associated with the environmental movement. Germany was a totally utilitarian society ... our view has changed very much and also the view on the river. When I was young, the rivers were polluted but people were fine with that. They thought we needed industry and pollution was just a side-effect. Later, people did not accept the side-effect anymore. (J. Freyhof at Leibniz Institute of Freshwater Ecology and Inland Fisheries, personal communication, 2 March 2018)
>
> We give nature more space. There'll be no change back to an old mentality because everybody realised water management is not only about flood protection, but also biodiversity and river is an important part of nature. I think people realised the importance of rivers earlier but 'where to begin' was the question. (B. Behm at Ministry of the Environment, Climate Protection and the Energy Sector, Baden-Württemberg, personal communication, 20 November 2017)

As a result, fewer dams were constructed after the 1970s when environmentalism increased in Germany. The idea of 'renaturing' rivers was promoted for many reasons: aesthetic, ecological, and practical advantages. Since the late 1970s, restoration projects have been pursued on major rivers, although the efforts were constrained due to already settled villages, fields, and industry installations (BMU/UBA, 2018).

5.5 Institutional Analysis

5.5.1 Federalism and the European Union (Formal Institutions)

The most prominent institutional feature of German politics is federalism. The geographic diversity of the country and individualism of the *Länder* based on historical background resulted in an unusual degree of fragmentation in German politics (Uekötter, 2014).

The federal government system has greatly influenced environmental policy. Historically, the legislation, management, and regulation of environmental policy including the water sector have been *Länder*'s responsibilities, and the federal government had limited power over the legislation of environmental policy (Schreurs & Papadakis, 2020; Bauer & Steurer, 2015). In 1972, a significant change occurred in the environmental protection system when the Constitution was amended. The federal government was granted concurrent legislative power, which meant that it was able to legislate in areas like air and noise pollution as well as waste management. The federal government could also release guidelines (framework laws) on water quality and nature preservation issues with implementation at the *Länder* level (Schreurs & Papadakis, 2020; Jänicke & Weidner, 1997). As there are different legal regulations at the federal and *Länder* levels, Jänicke and Weidner (1997) state that German environmental protection is based on one of the most complex legal systems in the world.

More dynamics were brought into German water management by EU environmental legislation. The importance of EU decisions has grown, especially in environmental fields as compared to other political agendas (Uekötter, 2014). In 1987, the Single European Act, which explicitly made the environment an area of EU competence, came into effect (Schreurs, 2002).

The EU offers a policymaking arena for tackling problems controversial in domestic politics (Wurzel, 2002). It provides legitimacy to environmental regulations that could face resistance from stakeholders. In the case of bathing water regulations, for example, the EU brought standards to Germany when the German government was not interested in the issue (Wurzel, 2002).

Federalism has sometimes hindered the implementation of European environmental policy in Germany. In 2000, the WFD came into force at the EU level, but the transposition of the directive into *Länder* water legislation lagged in Germany. The EC then sued Germany for insufficient implementation of the WFD (Court of Justice of the European Union, 2005). In turn, Germany carried out a major federal reform in 2006, and more detailed water legislation became possible at the federal level (UBA, 2013). Today, water legislation follows the 'competition principle' (*konkurrierende Gesetzgebungskompetenz des Bundes*). It means that *Länder* can establish water laws on their own only when the federal government has not entirely exhausted its national competence (BMU/UBA, 2018). When the federal government legislates a national and uniform law, this overrules the state-level laws (Metz, 2017). *Länder* can deviate from federal regulations, but it rarely occurs (BMU, 2016).

Länder are still responsible for the implementation of water legislation, and the federal government does not have supervisory powers. In *Länder*, water management is typically the task of the ministry of the environment, and there are sub-level authorities at district and municipality offices (BMU, 2016). Regarding EU regulations, the federal government represents Germany externally and it takes responsibility for implementing EU law, but the enforcement of EU law within Germany is a *Länder* task (UBA, 2013). Besides, *Länder* have their water acts (*Landeswassergesetz*) which supplement and transpose the federal laws (UBA, 2013).

The federal government and *Länder* meet up regularly in order to coordinate water management policy. The Working Group on water issues of the *Länder* and the federal government (*Bund/Länderarbeitsgemeinschaft Wasser*, LAWA) is comprised of representatives from the federal government and the *Länder*. The LAWA coordinates administrative implementation and legislation related to water management (BMU, 2016). Regarding the impacts of climate change on water resource management and potential adaptation options, a group of experts from the LAWA updates relevant documents and makes action recommendations (BMU/UBA, 2018).

Germany is also engaged in transboundary river management as it shares several rivers with other countries (Krysanova et al., 2010). Germany is currently a member of seven international commissions for transboundary river catchment areas, including the Rhine, Elbe, Danube, Oder, Maas, Moselle, Saar, and Lake Constance. After the WFD came into force, river basin management became obligatory among EU member states. Notably, the International Commission for the Protection of the Rhine Against Pollution (ICPR) established in 1950 shows a long tradition of transboundary management in the river basin (ICPR, 2015a; Krysanova et al., 2010).

Germany is active in international conventions related to water management as well. One of the conventions in which Germany is involved is the United Nations Economic Commission for Europe Convention on the Protection and Use of Transboundary Watercourses and International Lakes (UNECE Water Convention) and its Protocol on Water and

Health. Adopted in 1992, the convention aims to make a balance between water resource utilisation and ecosystem conservation/restoration. The activities of international river basin commissions including the ICPR and the ICPER are based on the principles of this convention (UBA, 2013). In 2000, the UNECE published guidelines on sustainable flood prevention emphasising preventive flood control and holistic river basin management (Economic Commission for Europe, 2000).

Germany also joined the UN Convention on the Law of the Non-Navigational Uses of International Watercourses. Germany is one of the few early signatories to the convention. The treaty provides a framework for governing shared freshwater resources in terms of equitable and reasonable utilisation and participation. Member states are obliged not to cause significant harm to other states which share an international watercourse (McCaffrey, 1999). Due to the high level of environmental protection and social responsibility required by the convention, it came into force in 2014, nearly 20 years after its adoption in 1997.

Table 5.2 Multi-level institutions regarding water management in Germany

	Organisations	Regulation/ guidelines
International	UNECE Water Convention UN Convention on the Law of the Non-Navigational Uses of International Watercourses	UNECE guidelines on sustainable flood prevention
EU	EEA	WFD Floods Directive
River basin commission	ICPER, ICPR, ICPDR, ICPO, ICPMS (International Commissions for the Protection of the Moselle and the Saar against Pollution), IMK (International Commission for the Protection of the Maas), and International Commission for the Protection of Lake Constance	
Federal	Ministry of Environment (BMU) UBA LAWA	Federal Water Act National Flood Protection Programme
Länder	State Ministry of Environment State environmental agency	State-level water legislations

Table 5.2 shows the complex structure of water management institutions in Germany. From the international conventions to *Länder* and municipal governments, relevant organisations and regulations play specific roles in the German water management system. Sometimes, such a distributed management system could hamper a policy change. According to Samuels et al. (2006), vertical as well as horizontal cooperation is a prerequisite to making any change in German policy. Informal processes to seek cooperation are, therefore, often more important than formal ones.

5.5.2 Participatory Governance (Formal and Informal Institution)

While federalism shows the vertical multi-level decision-making process of Germany, participatory governance shows the width of the multi-level process. In Germany, environmental policies are determined by cooperation between government and interest groups (Schreurs & Papadakis, 2020). This section explains how participatory governance became a formal, as well as an informal, institution.

Before participatory governance, it is worth noting that corporatism has worked as an informal institution in German politics, including environmental policy. Corporatism can be described as informal cooperation that exists among the government, bureaucracy, and opposition parties. Government officials found that corporatism could reduce resistance from different stakeholders. According to Uekötter (2014), corporatism was a defining feature in environmental policy in the twentieth century. Governmental officials delegated complex issues to committees that were comprised of experts and lobbyists. The access points during the discussion were relatively closed for external actors. The committees came up with a solution that would not face significant opposition from committee members, but environmental activists and the general public often remained largely excluded (Wurzel, 2002).

Uekötter (2014) points out that corporatism was an administrative practice particularly for environmental policy because environmental problems were seen as difficult issues that demanded expert knowledge.

> [Corporatism] in fact was the secret behind many achievements that have won Germany international acclaim [within the environmental sector]. The 'energy transition' towards renewables is based on corporatist alliances. (Uekötter, 2014, p. 4)

Until the early 1980s, corporatism obstructed the inclusion of new movements like the green party (*Die Grünen)* in the policymaking process. The dominant command-and-control style found in the environmental policy planning process, however, faced criticism for its inefficiency and inflexibility as pressure from environmental movements increased (Jänicke & Weidner, 1997). In turn, the federal government became more active in including *Die Grünen* in the informal process (Schreurs & Papadakis, 2020).

As *Die Grünen* entered parliament, the number of actors in the environmental policymaking process increased (Blühdorn et al. 1995). The decision-making process for environmental policy changed towards consensus-building and dialogue among diverse stakeholders (Schreurs & Papadakis, 2020; Jänicke & Weidner, 1997). The federal, state, and municipal governments started to ask for opinions regarding environmental issues from independent environmental organisations and research institutes. Jänicke and Weidner (1997) indicate that although the first independent research institute of Germany, ÖKO-Institut Freiburg, was established in 1977, it was only after the Chernobyl accident in 1986 that the Federal Ministry for Economic Affairs asked the institute about their opinions on its nuclear energy policy. Since more diverse actors started to be involved in the decision-making process, environmental politics became a more complex process which included both "contentious activities and cooperative decision making" (Schreurs, 2002, p. 21).

Now participatory governance is regarded as a compulsory element of German environmental policy. Most of the interviewees in Germany mentioned that one of the main characteristics of water management is the participatory process. It is not just a mandatory process by law, but the usefulness of participatory governance is recognised by the actors who are engaged in the policy process. According to interviewees, decisions made through consultations with stakeholders are seen as robust and acceptable.

> [In the policy process] Everyone has a chance to say what they want. If one person decides, you don't know whether it's correct or not. Many people have to think about [the issue] and come to the solution. Then the solution can be accepted by all, hopefully. It's better than one person's decision. [Our decision-making process] is in the middle of bottom-up and top-down, and this is good. (U. Hursie at Elbe River Basin Community Liaison Office (FGG Elbe), 23 November 2017)

Broadening participation has been pursued at the European level as well. According to Kaika (2003), multiple actors' involvement, power decentralisation, and increased concern for the environment were the main features of the WFD. In the policymaking process of the WFD, the EC invited water suppliers, chemical and fertiliser industries, agricultural organisations, farmers unions, environmental NGOs, and regulators (Kaika, 2004).

Participation in the WFD policy process meant consultation, not sharing the decision-making power (Harrison et al., 2001; Kaika, 2004). Nevertheless, environmental NGOs such as the Royal Society for the Protection of Birds, the European Environmental Bureau, and the WWF could influence the discussion to a certain extent and make the directive stricter for environmental protection (Kaika, 2004). Mainly, environmental NGOs were heavily engaged at the last stage of the WFD legislation when the European Parliament and the Council of Environmental Ministers of the EU had to reach an agreement (Kaika, 2004).

> [In the policymaking process of WFD,] NGOs had a chance to put [forward] some ideas. At that time, we had some very good people working as parliamentarians. The definition of 'heavily modified water bodies' or stricter regulations [are our achievements]. We discussed with other European NGOs about what we want and what we don't want. (M. Bender at Grüne Liga, personal communication, 16 November 2017)

In turn, the tendency towards the participatory approach at the EU level also influenced domestic politics in Germany. The WFD states that "member states should encourage the active involvement of all interested parties in the implementation of this directive, in particular in the production, review and updating of the river basin management plans" (Directive 2000/60/EC). Although the directive is criticised for its unclear indication of the participatory process, it provides a regulatory ground for it (Kaika, 2004).

Such participatory governance does not always result in a favourable policy for environmental protection. Accessibility to lobbying gained significant importance in the policymaking process, and the interest groups that have resources could be more advantageous than others to incorporate their agendas into the decision-making process (Kaika, 2004). In the WFD consultation process, for example, information regarding the policymaking process was not distributed enough to induce large-scale participation, although the open call had no access limitation. Therefore, the participation process was primarily influenced by visible groups of stakeholders who often had conflicting interests with each other (Kaika, 2004).

The formal and informal institutions analysed in this section show the complexity of multi-level governance within the German water management system. The institutions of water management have been expanded both in horizontal and vertical directions in Germany. The EU regulations on water became comprehensive and legally binding, adding more complexity to the German water management system based on federalism. Furthermore, participatory governance has been formalised by law and settled as a norm in the decision-making process of environmental politics. As can be seen in Sect. 5.4.2, environmentalism that grew along with environmental movements has been influential in such institutional changes. The interaction between discourse and institutions is further discussed in Chap. 6.

References

Alles harmlos. (1995, February 6). In *Der Spiegel* 6. Retrieved December 23 from https://www.spiegel.de/spiegel/print/d-9158530.html

Bauer, A., & Steurer, R. (2015). National Adaptation Strategies, What Else? Comparing Adaptation Mainstreaming in German and Dutch Water Management. *Regional Environmental Change, 15*(2), 341–352. https://doi.org/10.1007/s10113-014-0655-3

Bewerunge, L. (1988, March 31). Die Fluten fließen zu schnell. *Frankfurter Allgemeine Zeitung*, p. 10.

Björklund, G., Burchi, S., Connor, R., Cosgrove, W., Hendry, S., Moriarty, P., Rast, W., Salame, L., & Winpenny, J. (2009). Policies, Laws and Finance. In World Water Assessment Programme (Ed.), *The United Nations World Water Development Report 3: Water in a Changing World* (pp. 49–67). UNESCO & Earthscan.

Blackbourn, D. (2006). *The Conquest of Nature. Water, Landscape and the Making of Modern Germany*. Norton.

Blühdorn, I., Krause, F., & Scharf, T. (1995). Introduction: Getting Ready for the Third Phase. In I. Blühdorn, F. Krause, & T. Scharf (Eds.), *The Green Agenda. Environmental Politics and Policy in Germany* (pp. 11–24). Keele University Press.

BMU. (2005). *Act to Improve Preventive Flood Control*. Retrieved December 30, 2019, from https://www.bmu.de/en/law/act-to-improve-preventive-flood-control/

BMU. (2010). *Water Framework Directive: The Way Towards Healthy Waters*. BMU

BMU. (2016). *Water Protection Policy in Germany*. Retrieved May 1, 2020, from https://www.bmu.de/en/topics/water-waste-soil/water-management/policy-goals-and-instruments/water-protection-policy-in-germany/

BMU. (2017a). *Fragen und Antworten zum Nationalen Hochwasserschutz programm*. BMU. Retrieved December 28, 2019, from https://www.bmu.de/themen/wasser-abfall-boden/binnengewaesser/hochwasservorsorge/nationaler-hochwasserschutz/hochwasserschutzprogramm/

BMU. (2017b). *Kurzinfo Hochwasservorsorge und Hochwasserrisikomanagement*. BMU. Retrieved December 28, 2019, from https://www.bmu.de/faqs/nationales-hochwasserschutzprogramm/

BMU/UBA. (2018). *Water Resource Management in Germany. Fundamentals, Pressures, Measures*. UBA.

Bojanowski, A. (2013, June 11). Wie sich Deutschland besser schützen kann. *Spiegel Online*. Retrieved January 3, 2020, from https://www.spiegel.de/wissenschaft/technik/elbe-hochwasser-schutz-durch-polder-und-deiche-a-904894.html

Bölsche, J. (2002, October 3). Trittin Will "Stress für die Wälder senken". *Spiegel Online*. Retrieved December 23, 2019, from https://www.spiegel.de/politik/

deutschland/katastrophenschutz-5-trittin-will-stress-fuer-die-waelder-senken-a-216650.html

Brunotte, E., & Amberge, P. (2009). *Flussauen in Deutschland. Erfassung und Bewertung des Auenzustandes*. Bundesamt für Naturschutz.

Bubeck, P., Kreibich, H., Penning-Rowsell, E. C., Botzen, W. J. W., de Moel, H., & Klijn, F. (2015). Explaining Differences in Flood Management Approaches in Europe and in the USA – A Comparative Analysis. *Journal of Flood Risk Management, 10*(4), 436–445.

Bubeck, P., Klimmer, L., & Albrecht, J. (2016). Klimaanpassung in der rechtlichen Rahmensetzung des Bundes und Auswirkungen auf die Praxis im Raumordnungs-, Städtebau- und Wasserrecht. *NuR, 38*, 297–307. https://doi.org/10.1007/s10357-016-3003-1

Court of Justice of the European Union. (2005). *Vertragsverletzung eines Mitgliedstaats – Richtlinie 2000/60/EG – Gemeinschaftspolitik im Bereich der Wasserpolitik – Keine fristgerechte Umsetzung of 15/12/2005*.

Da-Cunha, L. V. (1989). Water Resources Situation and Management in the EEC. *Hydrogeologie, 2*, 57–69.

Deiche statt Harem. (1962, February 28). In *Der Spiegel* 9. Retrieved December 23, 2019, from https://www.spiegel.de/spiegel/print/d-45139169.html

Deutschländer, T., & Mächel, H. (2017). Temperatur inklusive Hitzewellen. In G. P. Brasseur, D. Jacob, & S. Schuck-Zöller (Eds.), *Klimawandel in Deutschland* (pp. 48–56). Springer.

Dromedar im Kanal. (1983, June 6). In *Der Spiegel* 23, Retrieved December 23, 2019, from https://www.spiegel.de/spiegel/print/d-14021530.html

Dryzek, J. S. (2013). *The Politics of the Earth. Environmental Discourses*. Oxford University Press.

Dryzek, J. S., Hunold, C., Schlosberg, D., Downes, D., & Hernes, H. (2002). Environmental Transformation of the State: The USA, Norway, Germany and the UK. *Political Studies, 50*(4), 659–682. https://doi.org/10.1111/1467-9248.00001

Economic Commission for Europe. (2000). *Sustainable Flood Prevention*. Retrieved April 2, 2019, from https://www.unece.org/fileadmin/DAM/env/water/publications/documents/guidelinesfloode.pdf

Ehrenstein, C. (2013, July 25). Deiche oder Auen? Das ist hier die Frage. *Welt*. Retrieved from https://www.welt.de/politik/deutschland/article118390294/Deiche-oder-Auen-Das-ist-hier-die-Frage.html

European Commission. (2000). *Directive 2000/60/EC of the European Parliament and of the Council of 23 October 2000 Establishing a Framework for Community Action in the Field of Water Policy*. Water Framework Directive.

European Commission. (2009). *Common Implementation Strategy for the Water Framework Directive (2000/60/EC). River Basin Management in a Changing Climate. EUR-OP (Technical Report, 40/2009)*. Retrieved May 1, 2020, from https://ec.europa.eu/environment/water/waterframework/facts_figures/guidance_docs_en.htm

European Commission. (2012). A Blueprint to Safeguard Europe's Water Resources. *COM, 2012*, 673.

European Commission. (2013). *An EU Strategy on Adaptation to Climate Change*. European Commission.

European Commission. (2016). *The EU Floods Directive*. Retrieved January 1, 2020, from https://ec.europa.eu/environment/water/flood_risk/index.htm

European Commission. (2018). *Report on the Implementation of the EU Strategy on Adaptation to Climate Change*. Retrieved December 4, 2019, from https://eur-lex.europa.eu/legal-content/EN/TXT/PDF/?uri=CELEX:52018DC0738&from=EN

European Commission. (2019). *Introduction to the EU Water Framework Directive*. Retrieved January 5, 2020, from https://ec.europa.eu/environment/water/water-framework/info/intro_en.htm

European Environment Agency. (2016). *Floods and Health*. Retrieved January 5, 2020, from https://www.eea.europa.eu/data-and-maps/indicators/floods-and-health-1/assessment

Federal Government. (2008). *Deutsche Anpassungsstrategie an den Klimawandel*. The Federal Government.

Federal Government. (2011). *Adaptation Action Plan of the German Strategy for Adaptation to Climate Change*. The Federal Government.

Federal Government. (2015). *Fortschrittsbericht zur Deutschen Anpassungsstrategie an den Klimawandel*. The Federal Government.

Federal Returning Officer. (n.d.). *Bundestag Election 1987*. Retrieved March 6, 2020, from https://www.bundeswahlleiter.de/en/bundestagswahlen/1987.html

Hairsine, K. (2017). The Wetlands Along the River Rhine Struggle to Recover from Human Engineering. In *DW*. Retrieved April 18, 2020, from https://p.dw.com/p/2WprY

Harrison, A., Schmidt, G., Avis, C., & Hauser, R. (2001). *WWF's Preliminary Comments on Public Participation in the Context of the Water Framework Directive and Integrated River Basin Management*. WWF.

Hirschfeld, J., Nilson, E. & Keil, F. (2014). *Everything Flows: A German Water Balance*. Edited by R. Harnisch. Retrieved January 5, 2020, from http://www.bmbf.wasserfluesse.de/?lang=EN

Hochwasser in den Flüssen zu einem Gutteil hausgemacht. (1980, November 2). In *Frankfurter Allgemeine Zeitung*, p. 9.

Huber, J. (2000). Towards Industrial Ecology: Sustainable Development as a Concept of Ecological Modernization. *Journal of Environmental Policy & Planning, 2*(4), 269–285. https://doi.org/10.1080/714038561

Hustedt, T. (2013). Analyzing Policy Advice: The Case of Climate Policy in Germany. *Central European Journal of Public Policy, 7*(1), 88–110.

ICPDR. (2013). *ICPDR Strategy on Adaptation to Climate Change*. International Commission for the Protection of the Danube River (ICPDR).

ICPDR. (2019). *Climate Change Adaptation Strategy*. ICPDR.

ICPER. (2015). *Internationaler Hochwasserrisikomanagementplan für die Flussgebietseinheit Elbe*. International Commission for the Protection of the Elbe River.

ICPO. (2015). *Hochwasserrisikomanagementplan für die Internationale Flussgebietseinheit Oder*. Retrieved January 9, 2020, from http://mkoo.pl/show.php?fid=5482&lang=DE

ICPR. (2015a). *History*. Retrieved January 3, 2020, from https://www.iksr.org/en/icpr/about-us/history/

ICPR. (2015b). *Strategy for the IRBD Rhine for Adapting to Climate Change*. ICPR.

Irgendwo mutt dat Woter jo hen. (1976, January 12). In *Der Spiegel* 3. Retrieved December 1, 2019, from https://www.spiegel.de/spiegel/print/d-41330655.html

Jänicke, M. (1985). *Preventive Environmental Policy as Ecological Modernisation and Structural Policy*. International Institute for Environment and Society.

Jänicke, M., & Weidner, H. (1997). Germany. In M. Jänicke, H. Weidner, & H. Jörgens (Eds.), *National Environmental Policies. A Comparative Study of Capacity-Building: With a Data Appendix, International Profiles of Change Since 1970* (pp. 133–156). Springer.

Kaczmarek, Z., Amell, N. W., & Stakhiv, E. Z. (1996). Water Resources Management. In R. T. Watson, M. C. Zinyowera, & R. H. Moss (Eds.), *Climate Change 1995. Impacts, Adaptations and Mitigation of Climate Change: Scientific-Technical Analyses. Contribution of Working Group II to the Second*

Assessment Report of the Intergovernmental Panel on Climate Change (pp. 469–486). Cambridge University Press.

Kaika, M. (2003). The Water Framework Directive: A New Directive for a Changing Social, Political and Economic European Framework. *European Planning Studies, 11*(3), 299–316. https://doi.org/10.1080/09654310303640

Kaika, M. (2004). Water for Europe: The Creation of the European Water Framework Directive. In J. Trottier & P. Slack (Eds.), *Managing Water Resources Past and Present* (pp. 89–116). Oxford University Press.

Kanalisierung versprochen. (1970, May 14). In *Frankfurter Allgemeine Zeitung*, p. 7.

Knauer, S. (1997, August 4). Auen statt Aufbau. In *Der Spiegel* 32, Retrieved December 23, 2019, from https://www.spiegel.de/spiegel/print/d-8755989.html

Knauer, S. (2004, June 11). Umwelt-Stiftung legt Auen-Institut trocken. In *Spiegel Online*, Retrieved December 23, 2019, from https://www.spiegel.de/wissenschaft/natur/hochwasserschutz-umwelt-stiftung-legt-auen-institut-trocken-a-303870.html

Krysanova, V., Dickens, C., Timmerman, J., Varela-Ortega, C., Schlüter, M., Roest, K., et al. (2010). Cross-Comparison of Climate Change Adaptation Strategies Across Large River Basins in Europe, Africa and Asia. *Water Resources Management, 24*(14), 4121–4160. https://doi.org/10.1007/s11269-010-9650-8

Künitzer, A. (2015, November 16). Water and the Green Economy: The EEA Perspective. In *11th European Water Association Conference*.

Kunz, M., Mohr, S., & Werner, P. C. (2017). Niederschlag. In G. P. Brasseur, D. Jacob, & S. Schuck-Zöller (Eds.), *Klimawandel in Deutschland* (Vol. 42, pp. 57–66). Springer.

Ländle unter. (1978, May 29). In *Der Spiegel* 22. Retrieved from https://www.spiegel.de/spiegel/print/d-40617885.html

Lücken im Deich. (1954, July 28). In *Der Spiegel* 31. Retrieved from https://www.spiegel.de/spiegel/print/d-28957058.html

Marx, S., Barbeito, G., Fleming, K., Petrovic, B., Pickl, S., Thieken, A., & Zeidler, M. (2017). *Synthesis Report on Disaster Risk Reduction and Climate Change Adaptation in Germany* (Vol. 56). DKKV-Schriftenreihe.

McCaffrey, S. C. (1999). *Convention on the Law of the Non-Navigational Uses of International Watercourses.* United Nations. Retrieved January 15, 2020, from https://legal.un.org/avl/ha/clnuiw/clnuiw.html

Metz, F. (2017). *From Network Structure to Policy Design in Water Protection. A Comparative Perspective on Micropollutants in the Rhine River Riparian Countries*. Springer.

OECD (2001). *OECD Environmental Performance Reviews:*

Pahl-Wostl, C. (2015). *Water Governance in the Face of Global Change*. Springer.

Radtke, I., Hustedt, T., & Klinnert, A. (2016). Inter-Ministerial Working Groups as a Panacea for Coordination Problems? *dms, 9*(1), 65–81.

Rucht, D. (1990). Campaigns, Skirmishes and Battles: Anti-Nuclear Movements in the USA, France and West Germany. *Industrial Crisis Quarterly, 4*(3), 193–222. https://doi.org/10.1177/108602669000400304

Samuels, P., Klijn, F., & Dijkman, J. (2006). An Analysis of the Current Practice of Policies on River Flood Risk Management in Different Countries. *Irrigation and Drainage, 55*(S1), S141–S150. https://doi.org/10.1002/ird.257

Schönthaler, K., & Andrian-Werburg, S. V. (2015). *Evaluation of the German Strategy for Adaption to Climate Change (DAS) – Reporting and Closing Indicator Gaps*. Umweltbundesamt.

Schreurs, M. (2002). *Environmental Politics in Japan, Germany, and the United States*. Cambridge University Press.

Schreurs, M., & Papadakis, E. (2020). *Historical Dictionary of the Green Movement*. Rowman & Littlefield Publishers.

Schwarz, K. (2015). *The National Flood Protection Program: Accelerating the Implementation of Large Scale Retention Measures within Germany*. BMU. Retrieved May 4, 2019, from https://circabc.europa.eu/webdav/CircaBC/env/wfd/Library/floods_programme_1/b_wg_f_on_floods/17th%20meeting%20-%2011-12_03_2015/Presentations/11c_National%20Flood%20Protection%20Program%20process%20to%20accelerate%20implementation%20of%20large%20scale%20retention%20measures.pdf

Stecker, R., Mohns, T., & Eisenack, K. (2012). Oldenburg Anpassung an den Klimawandel - Agenda Setting und Politikintegration in Deutschland. *ZfU, 2*, 179–208.

Steyaert, P., & Ollivier, G. (2007). The European Water Framework Directive: How Ecological Assumptions Frame Technical and Social Change. *Ecology and Society, 12*(1).

Thieken, A. H., Bessel, T., Kienzler, S., Kreibich, H., Müller, M., Pisi, S., & Schröter, K. (2016a). The Flood of June 2013 in Germany: How Much Do We Know About Its Impacts? *Natural Hazards and Earth System Sciences, 16*(6), 1519–1540. https://doi.org/10.5194/nhess-16-1519-2016

Thieken, A. H., Kienzler, S., Kreibich, H., Kuhlicke, C., Kunz, M., Mühr, B., et al. (2016b). Review of the Flood Risk Management System in Germany After the Major Flood in 2013. *E&S, 21*(2). https://doi.org/10.5751/ES-08547-210251

UBA. (2013). *Water Resource Management in Germany: Part 1, Fundamentals: Umweltbundesamt of Germany*. UBA.

UBA. (2019a). *Precautionary Principle*. Retrieved April 24, 2019, from https://www.umweltbundesamt.de/en/precautionary-principle

UBA. (2019b). *Competence center KomPass*. Retrieved December 30, 2019, from https://www.umweltbundesamt.de/en/topics/climate-energy/climate-impacts-adaptation/competencecenter-kompass

UBA. (2021). *Geplante Maßnahmen des Nationalen Hochwasserschutzprogramms sind wirksam*. Retrieved December 20, 2021, from https://www.umweltbundesamt.de/presse/pressemitteilungen/geplante-massnahmen-des-nationalen

Uekötter, F. (2014). *The Greenest Nation? A New History of German Environmentalism*. The MIT Press.

Weale, A. (1993). Ecological Modernisation and the Integration of European Environmental Policy. In J. D. Liefferink, P. Lowe, & A. P. J. Mol (Eds.), *European Integration and Environmental Policy*. Belhaven.

Wurzel, R. (2002). *Environmental Policy-Making in Britain, Germany and the European Union. The Europeanisation of Air and Water Pollution Control*. Manchester University Press.

6

Policy Change and Stability in Water Management in Face of Climate Change: A Comparison of South Korea and Germany

6.1 Discourse Comparison

6.1.1 Similarities and Differences in Social Discourse

In the previous two chapters, dominant social discourses related to the environmental policies of South Korea and Germany were examined based on the literature review and interview results. In South Korea, developmentalism, techno-centrism, and environmentalism were identified while optimism for technology and environmentalism were found in Germany. While these are general social discourses which are not particularly related to water management or flood risk management, the interviewees in both countries suggested that these general social discourses are some of the main drivers of policy change within the water management sector.

Techno-centrism in South Korea and optimism for technology in Germany are similar to each other. South Korean politicians have often made reference to the strong optimism for technology in Germany. Since the 1950s, when Germany had a high rate of economic growth and technological development, the South Korean government positioned

Y. h. Kang, *Climate Change Adaptation in River Management*, Palgrave Studies in Water Governance: Policy and Practice,
https://doi.org/10.1007/978-3-031-10486-2_6

Germany as its role model for economic growth. The South Korean concept of the Miracle on the Han River was borrowed from Germany's Miracle on the Rhine (*Wirtschaftswunder*) (Han, 2016).

The techno-centrism discourse of South Korea and optimism for technology of Germany have a similar attitude to nature. When optimism for technology was a dominant discourse in Germany, rivers were either viewed as a natural resource for economic development or as wild nature to be controlled (Blackbourn, 2006). When Korean President Lee Myung-bak attempted to convince the public of the benefits of the Grand Canal Project, he visited Germany to show that the canals on the Rhine and Danube rivers contributed to Germany's economic growth (Ahn, 2006).

The most significant difference between the two countries in terms of social discourse is the political power of environmentalism. Environmentalism first emerged as a robust social discourse in Germany in the 1980s, with environmental movements having been stronger in other industrialised countries such as the United States through the 1960s–1970s. The German environmental movement differed from others in that the political system helped environmental movements to enter the political party system. The German green party, *Die Grünen,* in Germany brought environmental issues to the political table and the party gained in popularity. In fear of losing votes to *Die Grünen*, other major parties then accepted environmentally friendly policies (Dryzek, 2013; Uekötter, 2014; Blühdorn et al., 1995). In this way, the environmentalism discourse in Germany extended its 'territory' into the political discursive space.

In South Korea, environmentalism has grown substantially in the last 30 years but is still not considered a major social discourse while developmentalism and techno-centrism remain powerful. A striking 78.6% of the respondents to a survey, reported by Kim et al. (2015), answered that they found the environmental condition of South Korea to be poor. However, environmental problems are viewed as an issue relating to individuals rather than structural problems. Willingness to participate in environmental movements through demonstrations, donations, and volunteering reduced between 2010 and 2014 (Kim et al., 2015). Kim et al. (2015) concluded that they did not observe much deep reflection on developmentalism from their survey results, although environmental awareness seems to increase gradually in South Korean society.

6.1.2 Programmatic Discourse Comparison

The prominent discourses in the water management sector were reviewed and categorised in Sect. 3.2.3 as 'engineering-development', 'participation-equity', and 'conservation-ecosystem service' approaches. These discourses form a triangular discursive space. Stakeholders in the water management field often try to convince other actors through the use of one of these discourses. In a democratic society, however, dominant programmatic discourse is influenced by various actors. Thus, the dominant discourse is usually located at a particular point within the discursive space formed by these three discourses.

Figure 6.1 shows where the dominant discourses on water management in South Korea and Germany are located in the discursive triangle. The leading discourse in South Korea is closer to an engineering-based development perspective, while the leading discourse in Germany has greater elements of participation-equity and conservation-ecosystem service. Although the programmatic discourses on water management both in Germany and in Korea were historically positioned in close proximity

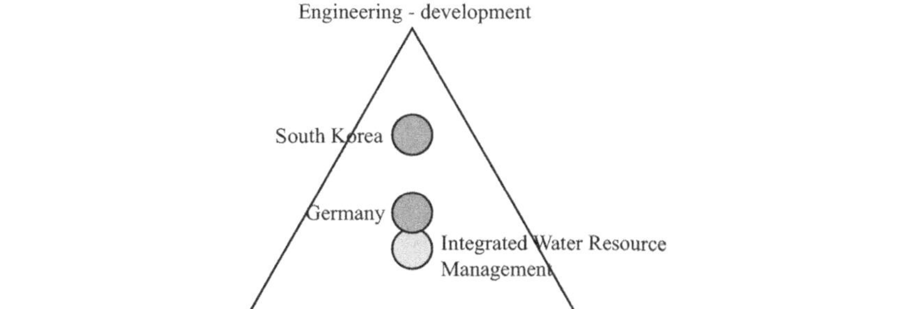

Fig. 6.1 The location of leading water management discourses in the water management discursive triangle

to engineering-based development, the current dominant discourses are distinctively different, as presented in their water policies. Such difference can be explained through the modern history of water management in the two countries.

The rivers in Germany have been heavily modified since the eighteenth century in order to make canals, generate electricity, prevent flooding and make use of the floodplains. The rivers were treated as an industrial resource for the purpose of economic growth (Blackbourn, 2006). For example, the Rhine 'correction' programme led by Johann Gottfried Tulla was a massive project to straighten the Upper Rhine River in the early nineteenth century (Hairsine, 2017). This Rhine River management programme influenced the development of the Han River in the 1960s in South Korea. Korean politicians regarded Germany's economic success after the Second World War as being linked, in part, to river engineering, and tried to follow Germany's growth model by developing the Han River (Lee, 1993; Prevent Disasters by Flood Control, 1962).

Strong environmentalism in Germany after the 1970s contributed to the shift in programmatic discourse towards a conservation-ecosystem service point of view. The environmental movement helped to expand the discursive space for stakeholders by opening up participation in decision-making processes and shifting the programmatic discourse closer to a participatory equity perspective. Another decisive factor in determining the programmatic discourse in Germany was the European-wide debate on integrated water resource management (IWRM). The Water Framework Directive (WFD) is referred to as an innovative regulation which institutionalised the idea of integrated water resource management through pro-ecological indicators and public participation (Steyaert & Ollivier, 2007; Kaika, 2003). Therefore, the leading programmatic discourse moved not only nearer to the conservation-ecosystem service perspective but also to greater participatory equity. As a result, discourses are now more balanced and supportive of IWRM.

In contrast, water management discourse in South Korea has not much moved away from the engineering-based development discourse that has dominated over the last 60 or more years. A policy document published by the Korean government in the 1970s stated that "rivers in their natural state cannot sustain our life. Without maintenance, rivers would only cause floods and drought" (Choi, 1971, p. 138). In Korean, the term 'river maintenance (하천정비)' connotes that rivers need to be controlled and modified. Concerns about water quality and environmental protection began to influence water policy in the late 1990s, particularly after chemical accidents on the Nakdong River, which caused drinking water contamination in the river basin (Ministry of Environment, 2010). In the divided institutional setting, however, the institutions which control water quantity have had more significant influence over water management policy than those dealing with water quality and ecosystem protection. The water quantity institutions are the MOLIT and K-water and they have been strongly biased by an engineering-based development discourse. The technical control of water bodies and infrastructure construction dominated the 'Long-Term Master Plan for Water Resource (2001–2020)', the main policy document for water management until 2018 (MOLIT, 2016).

6.1.3 Policy Discourse Comparison

Flood management is at the centre of climate change adaptation in water management, both in South Korea and Germany. Although the primary goal of flood management policy in both countries is to protect lives and reduce damage, the policy discourses in the two cases show differences in their approaches to the goal.

In South Korea, policy discourse related to flood management can be inferred from the discourse used to tout the Four Rivers Project. Civil engineering projects have been strongly associated with South Korea's fast economic development in the twentieth century, of which many Koreans are proud. As Zahariadis (2014, p. 37) puts it, symbols derived from the "core of a nation's identity" are more likely to be accepted in the policy process. As such, the government tried to frame the Four Rivers Project with rhetoric such as "landscape transformation through civil engineering" and "economic boost" (Ministry of Culture, Sports and Tourism, 2008; Kim, 2010). The storyline of the government's announcement of the Four Rivers Project can be summarised as 'this is the fundamental solution for preventing floods and droughts through an intensive civil engineering project, which can boost the economy'.

Another frame used for the policy discourse of the Four Rivers Project was that the project makes use of advanced technology. President Lee Myung-bak's talk during a TV debate showed that the Four Rivers Project was pursued based on a strong techno-centrism discourse.

> Lee: Those who are worried about water quality deterioration (due to the project) don't know about the current standard of Korea's technology. Korean technology on river restoration is world-class. How can a government in the 21st century make a plan to worsen water quality?
>
> Moderator: I heard that advanced technology would be applied to the project. Is that right?
>
> Lee: Yes. As you can see in this video, the robot fish will swim in the rivers and report water quality to the central monitoring system. After the project is completed in 2012, those robot fish will play an important role. (MBC, 2009)

Such a techno-centric frame was criticised by opponents who were concerned about the environmental impact of the project (e.g., Byun, 2011; Park, 2009). In response to the criticism, the government framed the project not just as a technical solution for river management, but also as an 'ecological restoration' project. Regardless of the appropriateness of the term for the project, the government's reaction shows that it could not entirely ignore a conservation-ecosystem service discourse. Nevertheless, the elements of the project were not changed during the design and implementation process and the engineering-development discourse predominated.

A public opinion survey on the Four Rivers Project conducted in 2010 shows the mixed and contradictory attitudes of the public towards the project (Lee, 2011). Although the respondents did not have high confidence in land development agencies' reliability, they had high expectations of national development projects. Forty-eight per cent of the respondents responded that the Four Rivers Project ignored public opinion and was a waste of money. As the construction of the Four Rivers Project progressed, however, opposition to the project decreased subsequently and supportive opinions increased. Local government leaders from the Democratic Party of Korea (opposition party at that time) changed their position from opposition to partial support (Realmeter, 2010). Interviewees pointed out that in the end, the opposition movement, firmly based on an ecosystem service and nature conservation discourse, could not earn enough support from the public to stop the Four Rivers Project.

In Germany, the National Flood Protection Programme (NFPP) adopted in 2014 is the primary flood policy. This policy was set up later than the German Strategy for Adaptation to Climate Change (DAS). The NFPP is in line with the water management section of the DAS because the European WFD and the Floods Directive provide the framework for both policies.

An underlying aim of the NFPP as outlined by the BMU is promoting "solidarity among *Länder* for flood risk management" (BMU, 2015). Traditionally, flood control was each *Länd*er's task, but this decentralised approach to flood management resulted in a lack of cooperation along river basins. Environmental NGOs including the German Federation for the Environment and Nature Conservation (BUND) criticised this structural fragmentation (Cornelius, 2014). Given the increasing scale of flood damages due to climate change and land-use change, the German government found it necessary to enhance a solidarity principle along the river basins. This principle emphasises balancing the burdens (flood risks) and benefits (flood protection) between upper stream and downstream areas (BMU, 2017). The NFPP is referred to as the first step to "softening the egoism" of *Länder* that cared only about their own lands within their borders (Cornelius, 2014).

A second goal of the NFPP is to implement preventive flood protection. Among other flood prevention strategies, polders which do not require heavy modification of water bodies can function as controlled retention areas when extreme floods occur. The Havel polder in Brandenburg showed the effectiveness of polders during the Danube and Elbe floods in 2002 (Förster et al., 2005). After the 2013 Danube and Elbe floods, large-scale retention measures gained more attention among policymakers and this tendency was reflected in the NFPP (Thieken et al., 2016). Rhetoric such as 'giving more space to the rivers (*den Flüssen mehr Raum geben*)' and 'synergies between flood protection and nature conservation (*Synergien für den Naturschutz*)' based on the ecosystem service approach can be found in the NFPP policy description (BMU, 2015). Because environmental friendliness is an essential criterion, projects for relocating dikes or constructing flood polders are prioritised within the NFPP (Schwarz, 2015; BMU, 2015).

When the frames, storylines, rhetoric, and metaphor of the two policies are compared, the NFPP is closer to the nature conservation and ecosystem service discourse than the Four Rivers Project. While technological advancement and economic development are emphasised in the discourse that supports the Four Rivers Project, such storylines and rhetoric cannot be found in the NFPP. Table 6.1 illustrates the summary of policy discourse comparison in regard to flood control in South Korea and Germany.

The external elements of the policy discourses differ in the two cases. First, as can be seen from the table, in both countries, some actors followed a discourse supportive of established policies while others opposed it. In South Korea, the main actors (including President Lee Myung-bak, the MOLIT, K-water, public research institutes, construction companies, and researchers with hydraulic and civil engineering backgrounds) formed a discourse-coalition in support of the Four Rivers Project. This coalition can be referred to as a development-oriented coalition that uses the frames and rhetoric such as 'an economic boost', 'green new deal', 'landscape re-creation', and 'high technology'.

On the other side of the discursive spectrum in South Korea, there is the environmental-protection coalition that consists of environmental activists from various NGOs (including the KFEM) and experts with backgrounds in environmental science or ecology. Approximately 2500 researchers organised the Professors' Organization for Movement against the Grand Korean Canal (POMAC) in order to stop the Four Rivers Project when the project was announced (POMAC, 2011). However, the power of the development-oriented coalition was strong; mainly due to support from the established institutions of the government. The developmental state institutions within the government structure represented by the strong presidency as well as the powerful MOLIT helped the development-oriented coalition dominate the decision-making process.

Table 6.1 A comparison of the main policy discourses on flood control

		South Korea (Four Rivers Project)	Germany (NFPP)
Internal elements	Frame	Landscape transformation through civil engineering	Solidarity among *Länder*; preventive flood protection
	Storyline	The fundamental solution for preventing floods and droughts is through intensive civil engineering projects, which can also help boost the economy	– A principle of solidarity among upper stream and downstream – Balancing burdens and benefits between upper stream and downstream areas based on the solidarity principle
	Rhetoric and metaphor	Boosting the economy, green new deal, landscape re-creation, high technology (robot fish as a symbol)	Giving more space to rivers, synergies between flood protection and nature conservation
External elements	Supporting actors	President, MOLIT, K-water, public research institutes, construction companies, researchers with hydraulic engineering and civil engineering backgrounds	BMU, UBA, EU, river basin committees, state governments, state environmental agencies, municipalities, Working Group of the Federal States on Water Issues (LAWA), environmental NGOs
	Opposing actors	Environmental NGOs, environmentalists, researchers with ecology and social science backgrounds	Farmers, residents in the river basins, industries that use the waterways
	Discourse-coalitions	Development-oriented coalition versus environment-protection coalition	Environment-protection coalition versus agriculture-industry coalition
	Links to institutions	Developmental state (strong MOLIT), strong presidency	Multi-level governance, EU directives (WFD, Floods Directive)

The findings of the German case are substantially different from that of South Korea. In contrast to the centralised decision-making process in South Korea, multi-level governance is a critical external element of the policy discourse for German flood management. The NFPP was developed following the EU Water Framework Directive and Floods Directive. In the course of the NFPP planning process, a number of actors at various levels such as the BMU, international and inter-state river basin committees, environmental NGOs, and *Länder* were involved. Notably, the Working Group of the Federal States on Water Issues (LAWA) plays an important role in German water policy.

The direction of flood risk management measures is the subject of debate between discourse-coalitions in Germany as well, but the difference from the South Korean case is that the environmental protection coalition is supporting the policy. While German environmental NGOs, ecologists, and many engineers promote more room for the rivers and such discourse is reflected in the policy, farmers, industry, and residents in the river basins have formed an opposing discourse-coalition. Farmers and residents who have lands in the floodplains often oppose floodplain restoration projects because they are concerned about their properties (Ehrenstein, 2013). Similarly, actors from industries and businesses that use rivers for shipping prefer hard infrastructure which stabilises the waterways. While the federal government tries to maintain the waterways, environmental NGOs argue that the cost-benefit ratio of the waterways needs to be examined and inefficient routes should be re-naturalised (Ehrenstein, 2013).

Different discourse-coalitions try to influence the decision-making process, resulting in policies that do not depend on a single discourse. The 2005 Flood Prevention Act and the NFPP have shown policy change to nature-based solutions. However, the president of NABU, a German environmental NGO, argued that language related to re-naturalisation in the 2005 Flood Prevention Act was "massively watered down" in the Bundesrat because of political interests. As a result, the ban on construction in floodplain areas can be exempted when the assessed flood risk is low according to the amended act (Ehrenstein, 2013). Another expert from BUND (a German environmental NGO) criticised that the NFPP still lacked a clear priority for ecological flood protection measures (Cornelius, 2014).

The external elements of the policy discourses explain why the Four Rivers Project and the NFPP were pursued in the respective cases. Even though South Korea and Germany are facing similar problems in their river basins due to climate change (i.e., more frequent and extreme floods), the differences in the power of their discourse-coalitions guided their policies in different directions. The development-oriented coalition in South Korea could push their frame in the Four Rivers Project with support from developmental state institutions and a strong presidency, while the environment-protection coalition in Germany could institutionalise their frame in the NFPP through various channels within the multi-level governance system. The differences in the formal and informal institutions which have influenced the flood risk management policies will be further explored in the next section.

6.2 Institutional Comparison

6.2.1 National Adaptation Policy

Similarities can be found in the development of climate change adaptation institutions in both countries. The first adaptation policies were established in 2008 both in South Korea and Germany, respectively, the National Climate Change Adaptation Master Plan of South Korea and the German Strategy for Adaptation to Climate Change (DAS). These adaptation policies are accompanied by rolling plans which are intended to be revised regularly. In South Korea, the supplementary policy is the National Climate Change Adaptation Policy (NCCAP) and in Germany, it is the Adaptation Action Plan. The leading ministry for the adaptation policy is the (Federal) Ministry of the Environment in both countries. Public research institutes such as the Korea Environment Institute (KEI) and the Federal Environmental Agency (UBA) of Germany provide scientific and technological support to the ministries' adaptation policies (Table 6.2).

Table 6.2 Adaptation policy and institutions comparison

	South Korea	Germany
Adaptation policy/main adaptation policy	National Climate Change Adaptation Master Plan in 2008 (for 2009–2030) (not legally binding)	German Strategy for Adaptation to Climate Change (DAS) in 2008 (15 action fields)
Supplementary policy	National Climate Change Adaptation Policy (NCCAP) (revised every five years) (10 fields)	The Adaptation Action Plan (regularly revised)
Leading ministry	– Ministry of Environment – Climate Change Adaptation Council	– Federal Ministry for the Environment, Nature Conservation and Nuclear Safety (BMU) – Permanent Committee On Adaptation to the Impact Of Climate Change (*Ständige Ausschuss Anpassung an die Folgen des Klimawandels*, AFK)
Leading research institute	Korea Adaptation Centre for Climate Change within the Korea Environment Institute (KEI)	KomPass within the Federal Environmental Agency (UBA)
Developed by	The KEI with experts from each field	The inter-ministerial working group in charge of adaptation policy (IWG Adaptation Strategy, *IMA Anpassungsstrategie*) prepared the Action Plan
Policy format	– Sectoral plans in detail – Compilation of policies from different ministries in the NCCAP; detailed planning and implementation of adaptation policy measures by each ministry or department	– Sectoral analysis and direction indication (no specific plans), indicating necessity of research and data collection – The action plan mainly deals with capacity-building activities including research, communication, framework, and standard-setting
Upper level/ background policy	Low Carbon Green Growth Act	Climate Protection Programme 2005 EU Adaptation Strategy

(*continued*)

Table 6.2 (continued)

	South Korea	Germany
Legislation sequence	From national to local level in consecutive order; central government's role—emphasised	Some Länder had sectoral or cross-sectoral adaptation strategies before the establishment of the DAS

The adaptation policy format of the two countries also shows similarities. The DAS and the NCCAP are a collection of separate ministries' and departments' plans. In the South Korean NCCAP, concrete actions by each ministry such as the development of a disaster management manual and vulnerability map are listed. It is a collection of sectoral plans. The first NCCAP (2011–2015) contains 87 detailed action plans formulated by 14 ministries. Similarly, the German DAS identifies 15 sectors for adaptation actions and indicates future policy direction, although it does not contain specific ministry plans.

These adaptation plans show the persistence of the institutions which were established before adaptation became a policy agenda. Climate change as a policy subject has not provided enough of a shock to society to restructure the whole institutional setting of a government. As discussed in Sect. 3.2.3, this is one of the challenges faced by adaptation policy in both South Korea and Germany. Notably, the path dependency of each sector hinders setting a coordinated adaptation policy. The way plans have been compiled into climate change adaptation plans means that sectoral policies are more weighted than the overall goal of the adaptation scheme. For instance, the German adaptation policy is planned by the inter-ministerial working group on adaptation policy (IWG Adaptation Strategy). Although the BMU prepares suggestions for the agenda and drafts working documents in the working group, its influence on respective issues of other ministries is limited (Radtke et al., 2016).

The two countries' national adaptation plans also have differences. First, the DAS refers to defining specific targets as the task of local or sectoral actors (BMU, 2008). In line with the DAS which emphasises the necessity of research and data collection, the Adaptation Action Plan mainly deals with capacity-building activities including research, communication, framework, and standard-setting. In contrast, the National

Climate Change Adaptation Master Plan lists a detailed plan and analysis by sector. Also, the NCCAP, the supplementary policy is a compilation of policies from different ministries. Each ministry and department involved in the establishment of the NCCAP provided detailed implementation plans for each ministry or department's adaptation measures.

Secondly, the actors who formulated the adaptation policies are slightly different. The adaptation policy of South Korea was mainly developed by the KEI, a governmental research institute under the Prime Minister's office. Sectoral specialists contributed to drafting the National Climate Change Adaptation Master Plan and the accompanying NCCAP (Ministry of Environment, 2008; Government of Korea, 2015). Government officers and experts designed detailed plans for implementing the measures. In the case of Germany, the DAS was formulated by an informal interdepartmental working group (BMU, 2014) and the Adaptation Action Plan was developed by IWG Adaptation Strategy. According to the UBA, almost all federal ministries participate in the working group, which is now the main institution of the German adaptation policy (UBA, 2019; A. Daschkeit, personal communication, 22 November 2017).

The legislative development sequence of the adaptation policies was also different between the two countries. Adaptation was first encouraged at the international level (e.g., the Least Developed Countries Work Programme of the UNFCCC). Many national governments started to establish national-level adaptation plans and subsequently set up sub-national adaptation plans (Fukumura et al., 2019; Abeysinghe et al., 2017). This was the case for the Korean adaptation policy. The discussion on adaptation started with the Korea Meteorological Administration and the Ministry of Environment, which were already leading the discussion on mitigation policy (Kwon et al., 2005). After the Low Carbon Green Growth Act came into effect, the NCCAP, provincial adaptation plans, and municipal adaptation plans were established consecutively (Government of Korea, 2015). In Germany, in contrast, the sub-national governments (*Länder*) started researching adaptation earlier than the federal government and before it established the national adaptation plan. Thirteen out of 16 *Länder* had already carried out studies on climate change impacts on different sectors and nine *Länder* developed climate change adaptation strategies (although most of them were sectoral, not cross-sectoral policies) before the DAS was established (BMU, 2008).

The last difference between the context within which the German and the South Korean adaptation plans were formulated, is the international level. In the German case, the discussions which took place within the supra-national European institutions were very important. The European Commission's *Green Paper on Adapting to Climate Change in Europe* (EC, 2007) and the *White Paper—Adapting to Climate Change: Towards a European Framework for Action* (EC, 2009) are mentioned in the DAS as influential papers shaping the German adaptation strategy. In the *Green Paper on Adapting to Climate Change in Europe*, for instance, soft non-structural measures that use natural processes for flood risk reduction were emphasised (EC, 2007). There was no comparable situation in South Korea as there is no formal regional organisation established in East Asia. Neighbouring countries such as Japan and China were still developing their adaptation plans when South Korea adopted its national adaptation plan in 2008 (Government of Japan, 2015; Tu et al., 2012).

Despite these differences, the institutional similarities in how climate adaptation policies affect the water management sector (particularly, flood risk management policy) appear to be greater. The similarity in the format of the adaptation plans, which compile policies from different departments and ministries, means that there is a persistence of sectoral policies in both countries. The national adaptation plans have provided little influence on existing policies, and consequently, the plans did not provide significant momentum for policy change in the water management sector.

6.2.2 Water Management Institutions

When the water management institutions of the two countries are compared, the differences are more evident than in the adaptation institutions (see Table 6.3). This is because the water sector institutions have been developed over a longer time than the adaptation institutions and have been influenced by discursive changes and critical events.

Table 6.3 Comparison of the formal water management institutions of South Korea and Germany with a focus on river management

	South Korea (until 2018)	Germany
Main policy	Long-Term Master Plan for Water Resource	Federal Water Act
Relevant law/ regulation	Long-Term Plan for Dam Construction Master Plan for Water Environment Management Integrated River Basin Flood Control Plan	EU Water Framework Directive EU Floods Directive
Leading Ministry/ Department	Quantity: Ministry of Land and Transport Quality: Ministry of Environment Small streams: Ministry of the Interior and Safety	Federal Ministry of the Environment, Nature Conservation and Nuclear Safety in Germany (BMU) (Traditionally, *Länder* are in charge of water management.)
Coordinating Organisation	None	Working Group of the Federal States on Water Issues (LAWA)
River basin management	River basin commissions (Han River, Nakdong River, Geum River, and Yeongsan River)	International and inter-regional river basin commissions (*Flussgebietsgemeinschaft*, FGGs)
Policy implementation agency	National and sub-national administration	Länder, international and FGGs

In South Korea, water quality and quantity management responsibilities have been divided between two ministries: The MOLIT and the Ministry of Environment. The fragmented water management system was established in the 1990s when a separate institution for environmental protection was created, and small streams were managed by the Ministry of the Interior and Safety. This system had been criticised by many scholars for its ineffectiveness (e.g., Ahn, 2012; Kim, 2013a; Chae, 2015), but it was not changed until 2018, when the newly elected government reformed water management institutions. Although the reform mainly aimed to integrate water quality and quantity management under the Ministry of Environment, the task of river management remained in the hands of the MOLIT leaving the integration incomplete (KFEM, 2018).

Another feature of the Korean water management institution is that the policies are mostly dealt with at the national level. The MOLIT's Long-term Master Plan for Water Resource was the central policy for water resource management until the 2018 reform. Other relevant policies include the Long-Term Plan for Dam Construction, the Master Plan for Water Environment Management, and the Integrated River Basin Flood Control Plan which were all established at the national level, mainly by the MOLIT. The 2018 reform was also pursued at the national level following President Moon's pledge (Government of Korea, 2017).

There have been attempts to decentralise river management in Korea. Regional commissions for managing water quality in the major rivers (including the Han, Nakdong, Geum, and Yeongsan) have been established since the late 1990s in order to facilitate consultation and mediation among residents and stakeholders. All the commissions are, however, chaired by the vice-president of the Ministry of Environment (Ministry of Environment, 2019). What is more, the senior manager of the water resource policy department in the MOLIT, the CEO of K-water, and the CEO of Korea Rural Community Corporation must be included in the commission by law. Although the heads of local governments in the river basins participate in the commission, they have limited authority in river management.

Water management in Germany is influenced by regulations at multiple levels. It is regarded as a 'joint task' of the federal (national), *Länder*, and municipal governments (BMU/UBA, 2018). While the central policy at the national level is the Federal Water Act, each *Länder* government has had its own water management policy. In addition, regulations at the EU level such as the Water Framework Directive and the Floods Directive must be followed by the federal and sub-national governments.

In order to coordinate water management policies, those who are responsible for water management policy in the federal government and the *Länder* participate in LAWA. In 2008, LAWA changed its policy discourse from flood control with hard infrastructure to flood risk management with a more integrated approach including soft infrastructure and re-naturalisation. In the policy document regarding the policy change

(*Strategie zur Umsetzung der Hochwasserriskomanagement-Richtlinie in Deutschland*), LAWA stressed that the change was built on the legal and technical input provided by the federal government and *Länder* (LAWA, 2008).

Another significant difference from the Korean water management institutions is that water management is regarded as the task of environmental ministries at both federal and *Länder* levels. Other federal ministries are also involved in water management, but their responsibilities are limited. For example, the Federal Ministry for Food and Agriculture promotes flow regulation and flood protection measures in the rural sector. Similarly, the Federal Ministry of Health is responsible for drinking water supply matters, as part of health care. Also, the Federal Ministry for Transport and Digital Infrastructure regulates federal waterways and navigation (BMU/UBA, 2018). Nevertheless, major decisions regarding water management are made by the BMU and state ministries of the environment.

6.2.3 Feedback, Interaction, Conflict: Informal Institutions of Decision-Making

The informal institutional features were pointed out both in South Korea and Germany when the interviewees were asked about the main feature of their water management policies. Although the difference between the water management policies cannot be solely attributed to the interaction and the feedback among actors, the interviewees' responses show that informal institutions are taken into account during their decision-making processes. This result is consistent with the perspective of new institutionalists who consider political institutions not as a mere reflection of society, but as "political actors in their own right" (March & Olsen, 1983, p. 738). Empirical studies since the 1990s continuously show that informal institutions contribute as independent variables in the policy process (Katzenstein, 1993; Polski & Ostrom, 1999).

In South Korea, the strong presidency and centralised decision-making process were commonly mentioned by many interviewees as the most significant informal institutional feature of Korean politics. Public officers conform to the president's will because they are afraid of disadvantages if they do not 'obey'. Although there is no formal regulation that restricts disobedience, public officers rarely protest against upper-level decisions. Their behaviour is justified by cases where public officers were disadvantaged after protesting. For example, a researcher at the Korean Institute of Civil Engineering and Building Technology was submitted to a severe disciplinary measure after he wrote on a website about his opposition to the Four Rivers Project. Two executive members of the labour union of the same institute were dismissed because they supported the researcher (Kim, 2013b). One interviewee described the situation within public research institutes during the Four Rivers Project as follows:

> Among the people who were directly involved in the Four Rivers Project, how many people thought the project was rational, do you think? I am not sure. Including me, public officers couldn't say anything publicly, but they must have thought it's nonsense. Nevertheless, the bureaucracy made the project possible. ... I think the governmental organisation system was the most crucial momentum. It is a closed system. Time was very limited. The [Four Rivers Project] master plan was completed within six months. In the top-down decision-making process, 'No' couldn't be accepted. ... When something is decided at the upper level, it is not rationally reviewed and improved at the lower level. Public research institutes try to justify the government's decisions. Those who opposed the Four Rivers Project got fired. So those who remained had to follow the plan, if they didn't want to get fired. In that case, what could you choose? (Anonymous, from Korea Institute of Civil Engineering and Building Technology, personal communication, 14 April 2017)

Although grassroots democracy was institutionalised in the general political system in the 1990s, local governments usually do not attempt to act against the national government's decision. One of the main reasons is the high rate of fiscal dependence. On average, 22.24% of the budget of local governments comes from the national government (Lim & Kum, 2018). The strong presidency is particularly influential in development projects for which the national government allocates most of the budget. For example, the MOLIT, the Ministry of Environment, and the Ministry of Agriculture and Food provided the budget for the Four Rivers Project (Board of Audit and Inspection, 2018).

Another feature of informal institutions in the South Korean policy process is that the decision-making process excludes external actors, including environmental NGOs. Even though a public hearing is obligatory in the process of planning and implementing a new policy, such a participatory process is often regarded as a formality. During the Four Rivers Project, President Lee Myung-bak expressed his thoughts about the opposition movement in a televised debate. He said that opposition is understandable but the opponents would eventually understand the merits of the Four Rivers Project once the project is completed (MBC, 2009). His speech is representative of the informal institution setting, which does not highly value participatory procedure and public discussions in the policymaking process.

> Lee: The Cheonggye-cheon project[1] faced enormous opposition from environmentalists, but after the completion, they supported it. I am familiar with opposition movements—I experienced this with the Cheonggye-cheon and the Gyeongbu Highway construction project. After the completion of the [Four Rivers] project, opponents will understand the necessity of the project. (MBC, 2009)

Informal institutions in the German policymaking process are substantially different from those of South Korea. The most frequently mentioned institutional feature during the interviews is the participatory process. Besides the fact that legal regulations require a participatory process, several interviewees find the participatory process necessary in environmental policies. They expressed their support for the participatory process because it could result in higher acceptance of policies. Indeed, the participatory process is believed to bring about more environmentally friendly outcomes.

[1] The Cheonggye-cheon project was an urban renewal project which restored a stream in the Seoul downtown when Lee Myung-bak was the mayor of Seoul between 2003 and 2005. The project was criticised by ecologists and environmental NGOs because the project did not reflect the fact that the Cheonggye-cheon was originally a dry stream. As a result, the restored stream is operated by an artificial system which intakes water from downstream (Lee, 2015).

> Everyone has a chance to say what they want. If one person decides, then you don't know if it's correct or not. Many people have to think about the problem and come to the solution. This solution will be accepted by all, hopefully. It's better than one person's decision. We take it in the middle of bottom-up and top-down and this is good. (U. Hursie from the Elbe River Basin Community Liaison Office, personal communication, 23 November 2017)
>
> When you involve stakeholders, there is a long participatory process. Then you cannot always come to the final goal—but in the end, it's more environmentally friendly, (Anonymous, geoecologist at a state environment agency in Germany, personal communication, 11 July 2017)

Public participation was not always an institutional feature of German environmental politics. It became increasingly institutionalised following the 1980s, as the environmental movement became more intense and *Die Grünen* entered into formal politics. Some scholars say that corporatism is still influential in German politics (e.g., Uekötter, 2014; Dryzek et al., 2002). The results from the interviews, however, indicate that the institution of water management has broadened from exclusive corporatism to a more open form with more diverse participants.

There is also criticism about the inefficiency of the participatory process in Germany. For example, the German flood risk management procedure is criticised because the involvement of the general public is often too late in the planning process (Thieken et al., 2016). The WFD requires stakeholders' participation in all water management policy processes but the implementation is not regulated by the EU. Experts suggest that more efforts to implement the early involvement of stakeholders are necessary (J. Hirschfeld, personal communication, 16 October 2017; Kaika, 2004).

The comparison of the two countries implies that informal institutions are a crucial factor influencing the different flood risk management policies. The flood risk management policies have been shaped through informal institutions such as participatory governance or centralised decision-making. The differences between the informal institutions of the two cases became more pronounced in the 1990s when discursive changes occur in Germany.

6.3 Policy Change and Stasis Through Discourse-Institution Interaction

6.3.1 Flood Risk Management Policy Comparison

This section compares the flood risk management policies of South Korea and Germany based on the discursive and institutional analyses in the previous two sections. Particularly, the focus is on the policy change and stasis that resulted in different policies: the Four Rivers Project (South Korea) and the National Flood Protection Programme (Germany). Some features of the two policies were already introduced in the previous sections which compared the water management and adaptation policies. Because two higher-level policies, namely adaptation and water management policies, are closely related to flood risk management, there are some overlapping features shared by policies at different levels. Nevertheless, this book distinguishes flood risk management, water management, and adaptation policies in order to trace the decision-making process systematically.

The governmental departments in charge of the flood risk management policies are different in South Korea and Germany, and this can be explained by their different institutional backgrounds. In South Korea, mainly developmental state institutions were involved in the Four Rivers Project. The master plan of the project was drafted by the Korean Institute of Civil Engineering and Building Technology which is a subsidiary research institute of the MOLIT. Subsequently, the project was implemented by K-water, a subsidiary institution of the MOLIT. In this process, the Ministry of Environment did not play a substantial role, although the project was expected to have a significant impact on aquatic ecosystems. The Ministry of Environment is the coordinating institution for the national adaptation policy (NCCAP), but the ministry could not take the lead role in the Four Rivers Project, which was instead the task of the MOLIT.

In contrast, the institutions for flood control in Germany are those in charge of environmental protection. Although the Basic Law states that flood protection is the *Länder*'s responsibility, a more coordinated policy became essential at the national level as flood risk has increased due to climate change. The NFPP, which came into force in 2014, is led by the Federal Ministry of Environment (BMU). This programme was developed by LAWA and the National/Federal Consortium for Nature Protection, Landscape Conservation and Regeneration (LANA) (BMU/UBA, 2018). The LAWA and LANA consist of representatives from federal and *Länder* ministries of environment.

The decision-making process also differs between the two cases. When President Lee initiated the Four Rivers Project, few state governors or mayors opposed the project (Realmeter, 2010). The post-Four Rivers Project plan such as the re-naturalisation of the rivers has been discussed at the national level, even after the ruling party changed in 2017. The flood control policy in South Korea can therefore be referred to as a typical top-down policy. In Germany, by contrast, the federal government plays a coordinative role, but it does not decide any specific projects for flood control. The federal government decided the NFPP's guidelines, but the projects under the NFPP are identified by *Länder* and submitted to the federal government.

Lastly, the main contents of the flood management policies differ. The Four Rivers Project was mostly focused on structural measures for flood protection: (1) constructing weirs, small and medium-sized dams, and reservoirs, (2) dredging river beds, (3) strengthening old dikes, (4) restoring ecosystems, and (5) creating bike lanes along the riverside (Ministry of Culture, Sports and Tourism, 2009). In contrast, the NFPP reflects the changed discourse for water management. The NFPP does not reject structural measures such as dikes and embankments, but it highlights non-structural measures that utilise floodplains and polders. Such policy contents of the NFPP are embodied by the phrase "give more space to the rivers" (Table 6.4) (BMU, 2017).

6.3.2 The Driver of Policy Change and Stasis

The differences between the flood management strategies of the two cases can be explained by policy change in Germany. As seen in Sect. 5.3.4, German flood risk management policy was similar to that of South Korea

Table 6.4 Main flood control policy of South Korea and Germany

	South Korea (before 2018)	Germany
Main flood control policy	Four Rivers Project River Act	The National Flood Protection Programme (NFPP)
Leading ministry/ department	Ministry of Land	Federal Ministry for the Environment
Developed by	Korean institute of civil engineering and building technology (Ministry of Land)	Working Group of the Federal States on Water Issues (LAWA), National/ Federal Consortium for Nature Protection, Landscape Conservation and Regeneration (LANA)
Implementation agency	K-water under the MOLIT	*Länder*, inter-regional river basin commissions (*Flussgebietsgemeinschaft*, FGGs)
Decision-making style	Top-down from national to local level President's order	Proposals submitted by *Länder* and checked by FGGs A guide from the national level
Relation to national adaptation policy	Included in the 1st National Climate Change Adaptation Policy	Established after the 1st DAS WFD and Floods Directive are commonly referred to within DAS and NFPP
Main contents	Constructing weirs, small and medium-sized dams, and reservoirs Dredging Strengthening old dikes Creating bike lanes = Structural means (engineering-based approach)	Giving more space to the rivers (regulated retention) Dike shifts Elimination of dike weak points = Mix of structural and non-structural means (mix of engineering and nature-based approaches)

until the 1990s. Germans reclaimed lands and canalised rivers for various purposes including flood protection from the mid-eighteenth century until the 1980s. River engineering projects in Germany were accelerated in the 1950s–1960s together with economic development in West Germany under the pressure of reconstruction (Blackbourn, 2006). Dike strengthening was repeatedly emphasised after severe floods (Lücken im Deich, 1954; Deiche statt Harem, 1962). Arguments for alternative solutions to flood protection, however, gained support in the policy process in the 1990s.

Many interviewees pointed out that raised environmental awareness was the main impetus behind policy change (Fig. 6.2). This driver was sometimes mentioned together with the environmental movement which started to spread widely in Germany in the 1970s. One interviewee argued that the changed attitude from utilitarianism to appreciating nature was another factor behind the flood risk management change. Thieken et al. (2016)) state that such discourse change signifies the move away from "purely technically oriented" flood defence measures.

Accidents and shocking events were also critical factors. When asked about the background of water policy change in Germany, some interviewees pointed out the Sandoz accident and the Chernobyl disaster. Although those accidents were not directly related to flood risk management, they were referred to as the triggers of water policy reform. This is because those accidents are linked to environmental awareness which was identified as the main impetus for policy change.

Floods were also mentioned as the main cause of policy change because they have provided opportunities to evaluate the existing policy's effectiveness and introduce a new approach (Thieken et al., 2016). For example, the LAWA proposed an integrated flood risk management for the first time after the floods in 1995 (LAWA, 1995). However, these

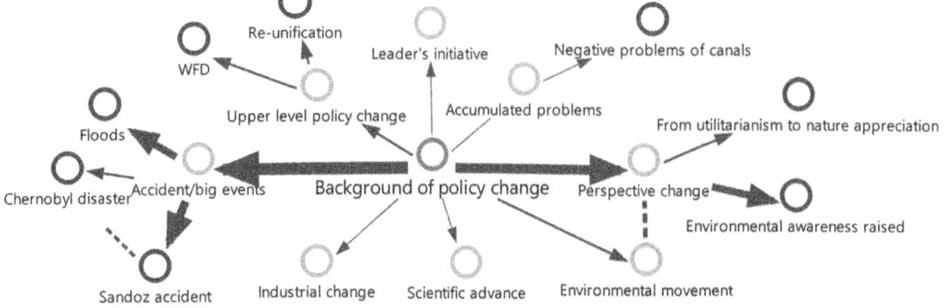

Fig. 6.2 Interview answers on the background of water policy change in Germany. Different colours of the circles indicate first (blue for the German case) and second (yellow) cycles of coding. The arrow's direction is from the latest coding to the earliest of the coding cycle

guidelines were not legally binding and they were not realised in action. The floods in 2002 were another shocking event in the German water management field. Due to the most severe flooding in at least 60 years, 21 people died and the damage totalled 11.6 billion EUR (Thieken et al., 2016). The German Flood Protection Act of 2005 (*Artikelgesetz zur Verbesserung des vorbeugenden Hochwasserschutzes*) embodies the policy changes made after the 2002 floods (Merz et al., 2014). Floods in 2013 again triggered changes in flood risk management policy, moving the discussion to large-scale retention measures. Accordingly, the NFPP was set up to identify potential areas for floodplains, dike relocation, and new polder areas (Thieken et al., 2016).

There were other drivers of policy change pointed out during the interviews. They included: (1) policy changes at higher levels of government, such as the WFD or political change like German re-unification, (2) accumulated environmental problems which made adverse problems visible, (3) leaders' initiatives, (4) industrial structural change, and (5) scientific advancement. Those factors were, however, less frequently mentioned than accidents/big events or perspective changes.

In contrast, the drivers of the Four Rivers Project indicate that substantial policy change has not occurred in the field of water management in South Korea. Engineering-development-based water management continued to dominate in the Korean policy process and resulted in policy stasis. As Fig. 6.3 shows, the two aspects of the project most frequently mentioned by the interviewees, were the government's strong drive and the expectation of development.

Several interviewees argued that the government pushed the Four Rivers Project because it was the president's plan. The project was not discussed enough among experts or the general public. Instead, scholars who raised their opposing voices felt pressure from the government to stop expressing their arguments (Anonymous 2 at Gyeonggi Research Institute, personal communication, 13 March 2017; Noh, 2012). The strong presidency together with stiff and top-down bureaucracy blocked differing opinions from entering into the decision-making process. The dotted lines in Fig. 6.3 represent the close links between these drivers of the Four Rivers Project.

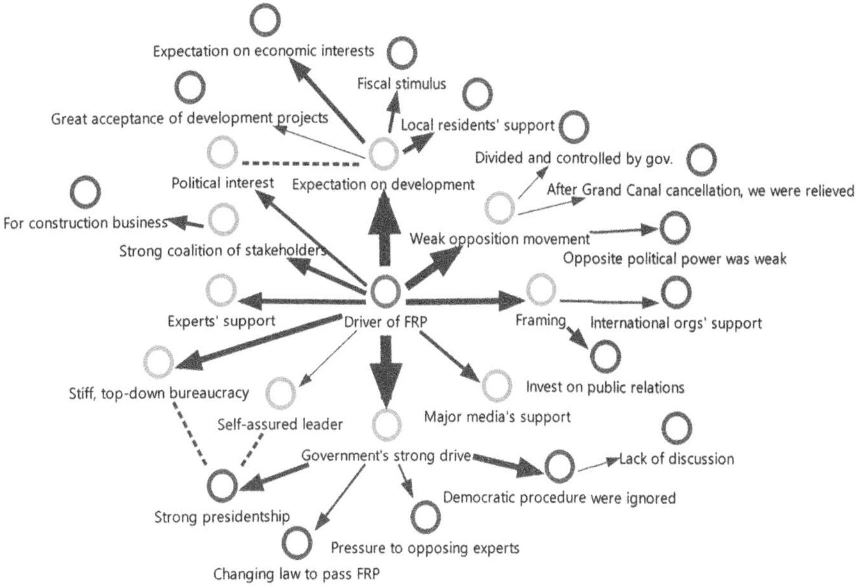

Fig. 6.3 Interview answers on the background of the Four Rivers Project in South Korea. Different colours of the circles indicate first (red for the Korean case) and second (yellow) cycles of coding. The arrow's direction is from the latest coding to the earliest of the coding cycle

The actors who could benefit from the Four Rivers Project formed a discourse-coalition. Particularly, it was clear from the planning phase of the project that construction companies would take advantage of the project. Many residents nearby the project sites welcomed it because they expected a rise in their real-estate value. Furthermore, major media were supportive of the project following the government's framing (Yun & Lee, 2010; Im, 2010). Indeed, the construction companies were able to support such major media by placing their advertisements in the media. This discourse-coalition was influential during the implementation of the project.

The high expectation the public placed on development projects was identified as another significant driver of the Four Rivers Project. The project formed the bulk of the Green New Deal policy, which was announced as a fiscal stimulus during the global recession in 2008. The components of the Four Rivers Project such as dredging and dam construction could be received by the public positively as a contribution to economic growth. Some interviewees, therefore, mentioned that the Four Rivers Project was pursued mainly out of the political interest of the president and the ruling party. Besides this, the opposition movement was not strong enough to mobilise the public. Many interviewees pointed out the opposition party's weak political power as well as the government's attempt to control public opinion, as the key factors behind this poor outcome.

6.3.3 Discourse Changes Influenced by Institutions

If both discourse and institutions are the drivers of policy change as well as policy stasis, how are they interrelated to each other? Was there an interaction or even a causal relationship between the dominant policy discourses and political institutions in the two cases? This section together with the next section examines this issue. First, this section looks into how discourse changes have been influenced by institutions in South Korea and Germany.

The flood risk management policies included in the national adaptation plans follow the programmatic discourse on water management. This is because the adaptation plans are a compilation of sectoral policies which allow path dependence in each sector. The policy discourses regarding flood risk management are therefore examined following the water management discourse categories that were analysed in Sect. 6.1.2.

Figure 6.4 illustrates the movement of the leading discourses in South Korea and Germany. The movement of the dominant discourse in South Korea was hindered by sticky institutions, which refer to the persistent developmental state and the strong presidency in this case. Although Germany also has persistent institutions such as federalism, Germany's leading discourse regarding water management moved away from the engineering-dominated development discourse. This is in part because of the multi-level governance setting it is embedded in and the multiple viewpoints this brought into debates.

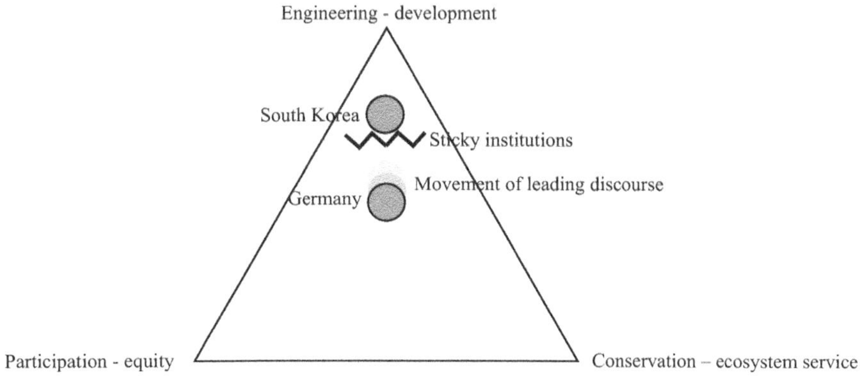

Fig. 6.4 The movement of the leading discourses in South Korea and Germany within the discursive space of water management

The findings from the case studies, particularly in the case of Germany, indicate that discursive change can lead to policy change. The power to alter the discourse comes from external shocks or internal accumulated energy such as scientific advancement and knowledge (Ländle unter, 1978; Alles harmlos, 1995). In the cases examined here, it can be observed that institutions either hinder or facilitate the movement of leading discourses within the discursive space. When institutions are sticky and in favour of vested interests, they hinder discursive changes in the society. In contrast, when institutions facilitate discursive flow among actors and they do not block the inflow of new ideas, discourse is more likely to change.

South Korea's flood risk management policy exemplifies the outcome of institutions that work against discursive changes. The strong developmental state institutions make it hard to move the dominant discourse away from the heavily engineering lead development focus towards other discourses. The rigid decision-making process, which does not easily allow the opposition to decisions made at higher levels, blocks the inflow of new ideas and discourses in the policy discursive space. As such, the Four Rivers Project could be realised, not because the rationale was very compelling, but because the leading discourse regarding water management (flood risk management) remained dominated by engineering perspectives.

The new government after 2017 tried to change water management institutions by transferring water quantity management tasks from the MOLIT to the Ministry of the Environment. The developmental state institutions including the MOLIT, K-water, and their allies in politics, however, have slowed the process down. While the government maintained that they would integrate all water management tasks into the Ministry of Environment, the task of river management was left to the MOLIT (Lim, 2018). Environmental NGOs and ecologists criticised the government's final decision to not completely transfer water quantity management tasks from the MOLIT to the Ministry of Environment (KFEM, 2018; Lim, 2018). In sum, the water management discourse in South Korea showed 'resistance' to changes, remaining close to the engineering-dominated development approach.

In contrast, the leading discourse regarding water management in Germany moved gradually from an engineering-heavy development focus to more participatory equity and conservation-ecosystem service orientations. The environmental movement post-1980s was an essential driver of the change within the water sector. Nevertheless, such a discourse transition would not have been smooth without the particular institutional features of German environmental politics. Especially, the multi-level governance institutions, reaching from the EU level through to the municipal level, allowed diverse ideas and discourses to flow into politics.

Even in Germany, the engineering-dominated development approach is still preferred by many stakeholders such as farmers and local residents, according to several interviewees. This support hinders more radical change away from the dominant discourse towards the more conservation-ecosystem service perspective. Conflicts among discourse-coalitions have made fundamental policy changes difficult in the European water management framework. The conflict between engineering and nature-based solutions is an ongoing issue at all government levels from the EU down to municipalities.

Pelling's (2011) three types of adaptation described in Sect. 3.1.5 are highly relevant for evaluating the movements in the leading discourses of water management in South Korea and Germany. This is because the water management parts of the national adaptation plans in both cases follow the sectoral policies. Pelling identifies different types of adaptation based on the scope and range of change: resistance, transition, and transformation. In South Korea's case, the discourse's movement can be described as a resistance to new challenges, while the leading discourse in Germany showed a transitional movement. The different discourse movements of the two countries led to different policies for flood risk management. Therefore, the different patterns of discursive change examined in the case studies imply that institutions can be a determining factor in different types of adaptation policy.

To conclude this section, institutions can be both a transformation facilitator and a sticky force to discursive changes. It is important to note that the movement of programmatic and policy discourses does not inevitably occur following a social discourse change or solely due to institutions. Schreurs (2002) suggests that a strong environmental movement does not always bring strict environmental regulations. This argument is consistent with many interviewees' responses which emphasised the critical role of shocking events, such as the Sandoz accident or severe floods, in contributing to discourse change in Germany.

6.3.4 Path Dependency and Critical Junctures Influenced by Discourse[2]

The previous section analysed how institutions influence the movement of discourses in the water management discursive space. But what about the opposite direction? This section examines how discourses impact the institutional stability and change in order to investigate the relationship between discourses and institutions in more depth.

In South Korea, formal institutions that manage water quantity have hardly changed over time. When chemical accidents (i.e., phenol and organic solvent leakage) occurred in the early 1990s, environmental issues gained attention in South Korean society and the Ministry of the Environment was established for regulating water quality. Environmental protection regulations such as environmental impact assessments and

[2] Ideas from an early version of this section were presented at Adaptation Futures 2018 on 19 June 2018 and appeared in the conference paper.

strategic impact assessments were also introduced (Ministry of Environment, 2010). Nevertheless, water quantity institutions represented by the MOLIT and K-water have maintained their tasks including dam development and river maintenance.

Compared to formal institutions, informal institutions have shown even more resistance to change. This is likely due to the fact that the socially constructed understanding of advantage and disadvantage cannot be easily altered. The informal institutions formed during the period of fast economic development in the 1960s–1970s are still powerful. These included a strong presidency and a developmental state structure with close relationships between government bureaucracy (i.e., land development agencies including K-water) and big corporations (i.e., construction companies). The developmental state institutions have exerted decisive power, particularly in regard to development projects.

Some interviewees in South Korea pointed out that environmental protection procedures were ignored during the planning and implementation process of the Four Rivers Project because of the relevant actors' strong belief that the economic benefits of the project were larger than that of conservation. Environmental NGOs and some local residents along the rivers sued the Korean government for not conforming to procedural regulations, but all the lawsuits were lost (Y. Lee, personal communication, 6 March 2017). Interviewees argued that development projects have hardly been challenged by the National Assembly or by the judiciary, and the reason is that developmentalism is stronger than environmentalism in society. Although the number of environmental NGOs has increased and their discursive power has grown significantly since the 1990s, the interviewees who have been involved in environmental movements in South Korea had negative outlooks on environmentalism in the near future.

> Environmental lawsuits are 100% a losing game. ... The judges were not enthusiastic [about the Four Rivers Project case] during the trials. They follow power. The Chief Justice of the Supreme Court is appointed by the president. In this structure, they can't be independent from the presidential power. (Y. Lee, lawyer who was in charge of the Four Rivers Project lawsuit cases, 6 March 2017)

In Germany, the interviewees, including public officers, expressed their opinion that the participatory process is necessary, although they do not always find it 'enjoyable'. Their answers imply that the participatory process is now recognised not just as a compulsory step by law, but also as an informal institution which is expected to bring certain incentives if followed (Helmke & Levitsky, 2004). Interviewees mentioned that participatory processes could contribute to making policy more robust.

The changes in formal and informal institutions in Germany have occurred along with a change in social discourse tied to the growth of environmentalism since the 1980s. The actors of the environmental movement started to demand stricter environmental regulations and more involvement in decision-making procedures. The popularity of *Die Grünen* showed the growing power of environmentalism. Considering that more recently all the political parties claim that environmental protection has become a priority for them, the BMU is now one of the most powerful of such ministries in the world (Schreurs, 2002).

Meanwhile, shocking events have not always led to substantial policy change in Germany. Several chemical accidents were not reported widely before the Sandoz chemical spill, hardly resulting in institutional changes. Before the Sandoz chemical spill, there were already suspicions that poisonous materials had been discharged into European rivers. For example, a spill of highly poisonous pesticides into the Rhine resulted in the mass death of fish in 1969. The cause of the accident remained "a mystery" at that time (Tümmers, 1994). The Federal Health Office launched an investigation and identified Hoechst AG but the company denied their responsibility (Wir sollten aufwachen und überlegen, 1986). In 1979, the same company spilt chemicals again into the Main River. It was reported that Werner Brans, State Secretary of the Hessian Ministry of the Environment, admitted that he had known "for several months" that Hoechst was discharging sewage directly into the Main (Koch, 1979). Also, 25 incidents and accidents in chemical plants occurred in the greater Cologne area between 1980 and 1986 (AK Chemische Industrie and KATALYSE, 1986, as cited in Wir sollten aufwachen und überlegen, 1986). Such incidents occurred repeatedly but they did not result in a major change in water management policy (critical juncture) until the Sandoz chemical spill in 1986 when environmental awareness was high after the Chernobyl disaster (A. Schulte-Wülwer-Leidig at ICPR, personal communication, 27 October 2017).

Multi-level governance was also referred to by the interviewees as a major factor behind changes in national and sub-national water management regulations. Without lobbying from environmental NGOs and lobbyists, the changes in the EU regulation, particularly in the water management sector, could have been much slower (Kaika, 2004; Geiler, 2001). However, it is important to note that multi-level governance itself does not bring policy change as can be seen in events pre-1990. Before 1986, the federal government tried to control water quality by amending the Federal Water Act (e.g., the fourth amendment in 1976), but the impact of such attempts was limited as water management has been *Länder*'s responsibility (Koch, 1979). This implies that change in policy discourses was a necessary condition for the policy change in Germany.

The two cases of this study provide evidence that discourses affect institutional stability as well as change. When the dominant discourse is strong enough to reinforce existing institutions, it is hard to expect new institutions to grow (both informal and formal institutions). Even if shocking events occur, they cannot make a substantial change in institutions when the dominant discourse does not support such a change. When the leading discourse changes through the emergence of alternative discourse, existing institutions can be weakened by such a change. Figure 6.5 is a schematic framework which explains how discourses influence institutional change in the policy process. The arrows represent the power of discourses that either suppress or strengthen the emergence of new institutions (including policy).

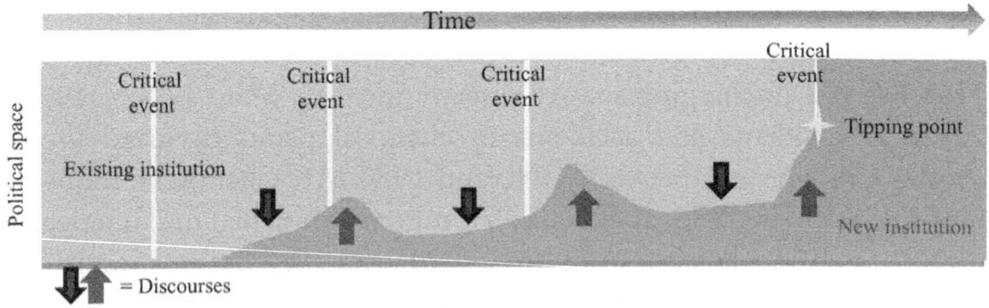

Fig. 6.5 Discourse-institution policy process framework

This study indicates that if a shocking event occurs concurrently with a change in discursive dynamics, policy changes can be brought forth. Although the policy stability created by the "tight connection between institution and idea" (Baumgartner & Jones, 2009, p. 16) can last for a long time, it can be broken once the definition of issues (i.e., discourse) changes. The point where the increasing force of alternative discourse meets a major event (critical juncture) can become the tipping point for policy change (Kang, 2019). When the political dynamics go beyond the tipping point, new policies or institutions are often planned and implemented.

6.3.5 Confirming and Contributing to Discursive Institutionalism

The results of this book are consistent with the earlier findings of the scholars of discursive institutionalism. As mentioned in Sect. 2.1.2, discursive institutionalism is an umbrella concept which accounts for the interactive processes of discourses. It acknowledges the significant role of institutional circumstances in which the ideas are located (Schmidt, 2009, 2010). However, institutions may change when discursive change occurs because people are able to think outside of institutional frames (Schmidt, 2015). Discourse and institutions are both the "structures and constructs" of one other (Giddens, 1990, as cited in Palmer, 2010, p. 997). In the perspective of discursive institutionalism, policy changes not only through endogenous processes through discursive interactions, but also with exogenous processes such as shocking events (Schmidt, 2010).

This study found that discourse plays a role in enabling or constraining institutions, and this result is in agreement with Schmidt's (2008) argument. The findings of this book also support Hall's (1993) policy paradigm theory, which argues that the change in the economic policy in Britain in the 1970s was the result of an evolving societal debate. While Hall suggests a quite linear process of events, crisis, and, subsequently, the new paradigm's emergence, this study argues that discursive interactions are often messy. Even if a crisis occurs, policy change may not happen when the prevalent discourse continues to support the existing institutions. The situation in Germany before the rise of environmentalism in the 1980s shows this. Severe chemical accidents before the Sandoz spill did not lead to an immediate reaction among the public or politicians.

The case studies show that both formal and informal institutions work either as a driving force to facilitate or restrict discursive change, and this finding accords with earlier observations, which showed that institutional logics can restrain policy transformation (Bosomworth, 2018; Schmidt, 2010). The difference between South Korea's and Germany's flood risk management was explained by policy change in Germany, and discursive change following the increasing power of the environmental movement was identified as the main reason for this policy change. Multi-level governance facilitated the alternative discourse's inflow to the decision-making process. In contrast, the rigid developmental state institutions together with the strong presidency in South Korea have hindered the entry of environmentalism into the decision-making process. The differences between the two cases explain the significant role of institutions in the degree of discursive change. When Pelling's (2011) three types of adaptation (resistance, transition, and transformation) are brought into the discussion, institutions seem to play the key role in determining the type of discourse changes.

The results of this study also indicate that the windows of opportunity for water management provided by events such as chemical accidents and floods, do not always lead to institutional change. The difference in the policy responses to critical events comes from the differences in dominant policy discourses and institutional settings. Historical institutionalists argue that critical junctures are the point of institutional change and they are usually triggered by external shocks. However, this study found that shocking events are not enough to change the existing, self-reinforcing institutions. If there are no alternative policy proposals available in the political space, shocks do not lead to institutional changes including policy changes. Existing institutions can quickly stabilise the situation and maintain the status quo with supportive discourses. As Samuels et al. (2006) emphasise, "a disaster makes progress" when "the seeds for such a change" are already spread (p. S149).

Taking Schmidt's framework (Schmidt, 2008, 2010) one step further, this study adds to the expanding field of discursive institutionalism in environmental policy. First, it suggests a discursive space where the leading policy discourse may be located. Although the triangular discursive space was applied only to the policy discourse related to water management in this book, the concept of discursive space can be extended to other fields of policy. Second, this study combined the three types of adaptation suggested by Pelling (2011) to the discursive change and interaction. Lastly, this study suggests a visualised discourse-institution policy process framework as presented in Sect. 6.3.4. Using the concept of the tipping point, this study tries to explain how policy change occurs through the interaction of discourses and institutions.

References

Abeysinghe, A., Dambacher, B. M., & Byrnes, R. (2017). *National Adaptation Plans: Understanding Mandates and Sharing Experiences*. IIED.

Ahn, S. (2006, October 25). Lee Myung-bak Visited Rhine-Donau Canal. *KBS*. Retrieved from http://news.kbs.co.kr/news/view.do?ncd=1241148

Ahn, K. S. (2012). A Study on Improvement of Overlapping Functions of the National Water Management Policy. *Korean Society and Public Administration, 23*(3), 183–208.

AK Chemische Industrie & KATALYSE. (1986). *Gift macht Geld. Die chemische Industrie und Strategien zu ihrer Entgiftung* (1st ed.). Kölner Volksblatt.

Alles harmlos. (1995, February 6). In *Der Spiegel* 6. Retrieved December 23, 1995, from https://www.spiegel.de/spiegel/print/d-9158530.html

Baumgartner, F. R., & Jones, B. D. (2009). *Agendas and Instability in American Politics* (2nd ed.). University of Chicago Press.

Blackbourn, D. (2006). *The Conquest of Nature. Water, Landscape and the Making of modern Germany*.

Blühdorn, I., Krause, F., & Scharf, T. (1995). Introduction: Getting Ready for the Third Phase. In I. Blühdorn, F. Krause, & T. Scharf (Eds.), *The Green Agenda. Environmental Politics and Policy in Germany* (pp. 11–24). Keele University Press.

BMU. (2008). *Deutsche Anpassungsstrategie an den Klimawandel: Hintergrundpapier*. Retrieved July 8, 2019, from https://www.bmu.de/download/deutsche-anpassungsstrategie-an-den-klimawandel/

BMU. (2014). *Anpassung an den Klimawandel*. Retrieved December 31, 2019, from https://www.bmu.de/themen/klima-energie/klimaschutz/anpassung-an-den-klimawandel/

BMU. (2015). *Nationales Hochwasserschutzprogramm*. Retrieved January 6, 2020, from https://www.bmu.de/faqs/nationales-hochwasserschutzprogramm/

BMU. (2017). *Kurzinfo Hochwasservorsorge und Hochwasserrisikomanagement*. Edited by BMU. Retrieved December 28, 2019, from https://www.bmu.de/faqs/nationales-hochwasserschutzprogramm/

BMU/UBA. (2018). *Water Resource Management in Germany. Fundamentals, Pressures, Measures.*

Board of Audit and Inspection. (2018). *An Analysis of the Implementation and Outcome of the Four Rivers Project*. Board of Audit and Inspection. Retrieved December 26, 2019, from http://www.bai.go.kr/bai/groupreport/auditResults.do?mdex=bai222&rpstOrgCd=0000000024&audYr2=2017&audNo=509&dspRqSno=10&isSearch=1

Bosomworth, K. (2018). A Discursive–Institutional Perspective on Transformative Governance: A Case from a Fire Management Policy Sector. *Environmental Policy and Governance, 28*(6), 415–425. https://doi.org/10.1002/eet.1806

Byun, D.-g. (2011). A Review on the Four Rivers Restoration Project from the Viewpoint of the Collaborative Governance Systems for the Sustainable Development in Korea. *Journal of Social Science, 23*(2), 31–63.

Chae, Y. G. (2015). A Study Concerning Integrated Water Management System in Korea. *Environmental Law Review, 37*(3), 309–352.

Choi, D. (1971). Government's Policy Direction (1). Commitment to Prosperity Which Will Change the Map—Comprehensive Plan for Developing the Four Rivers. *Korean Regional Administration Society, 20*(207), 138–149.

Cornelius, C. (2014, October 22). Wer zahlt die Deiche? *Der Spiegel*. Retrieved from https://www.spiegel.de/wissenschaft/natur/hochwasser-schutz-bund-und-laender-wer-zahlt-fuer-investitionen-a-998589.html

Deiche statt Harem. (1962, February 28). In *Der Spiegel* 9, Retrieved December 23, 2019, from https://www.spiegel.de/spiegel/print/d-45139169.html

Dryzek, J. S. (2013). *The Politics of the Earth. Environmental Discourses*. Oxford University Press.

Dryzek, J. S., Hunold, C., Schlosberg, D., Downes, D., & Hernes, H. (2002). Environmental Transformation of the State: The USA, Norway, Germany and the UK. *Political Studies, 50*(4), 659–682. https://doi.org/10.1111/1467-9248.00001

Ehrenstein, C. (2013, July 25). Deiche oder Auen? Das ist hier die Frage. *Welt*. Retrieved from https://www.welt.de/politik/deutschland/article118390294/Deiche-oder-Auen-Das-ist-hier-die-Frage.html

European Commission. (2007). *Directive 2007/60/EC of the European Parliament and of the Council of 23 October 2007 on the Assessment and Management of Flood Risks (Floods Directive)*

European Commission. (2009). *White Paper: Adapting to Climate Change: Towards a European Framework for Action*

Förster, S., Kneis, D., Gocht, M., & Bronstert, A. (2005). Flood Risk Reduction by the Use of Retention Areas at the Elbe River. *International Journal of River Basin Management, 3*(1), 21–29. https://doi.org/10.1080/15715124.2005.9635242

Fukumura, Y., Kazutaka, O. & Hijioka, Y. (2019, May 28). National Adaptation Policy and Its Influence on Local Practice: Climate Change Adaptation Act, Japan. In *4th European Climate Change Adaptation Conference*. ECCA. Retrieved April 18, 2020, from https://www.ecca2019.eu/national-adaptation-policy-and-its-influence-on-local-practice-climate-change-adaptation-act-japan/

Geiler, N. (2001). The Role of the Citizens' Action Committees in Water Protection Policies. In K. Rudolph (Ed.), *The German Water Sector. Policies and Experiences* (pp. 34–42). Federal Ministry for the Environment, Nature Conservation and Nuclear Safety.

Giddens, A. (1990). *The Consequences of Modernity*. Polity Press.

Government of Japan. (2015). *National Plan for Adaptation to the Impacts of Climate Change*. Retrieved April 15, 2020, from http://www.env.go.jp/en/focus/docs/files/20151127-101.pdf

Government of Korea. (2015). *The 2nd National Climate Change Adaptation Plan (2016-2020)*. Government of Korea.

Government of Korea. (2017). *100 Major Governmental Projects*. Retrieved January 12, 2020, from http://www.opm.go.kr/_res/opm/etc/kukjungfile.pdf

Hairsine, K. (2017). The Wetlands Along the River Rhine Struggle to Recover from Human Engineering. In *DW*. Retrieved April 18, 2020, from https://p.dw.com/p/2WprY

Hall, P. A. (1993). Policy Paradigms, Social Learning, and the State: The Case of Economic Policymaking in Britain. *Comparative Politics, 25*(3), 275–296.

Han, Y. (2016). *Miracle from the Rhein to the Han River: Heavy industrialisation of South Korea and Its Social Consequences Under Park Chung Hee.* PhD. Lancaster University. Retrieved October 16, 2019, from https://eprints.lancs.ac.uk/id/eprint/79039/

Helmke, G., & Levitsky, S. (2004). Informal Institutions and Comparative Politics: A Research Agenda. *Perspectives on Politics, 2*(4), 725–740.

Im, Y.-J. (2010). A Comparative Analysis of News Frame Based on the Public Enterprise: The Grand Canal in the Korean Peninsula. *Korean Journal of Communication & Information, 49*(1), 57–80.

Kaika, M. (2003). The Water Framework Directive: A New Directive for a Changing Social, Political and Economic European Framework. *European Planning Studies, 11*(3), 299–316. https://doi.org/10.1080/09654310303640

Kaika, M. (2004). Water for Europe: The Creation of the European Water Framework Directive. In J. Trottier & P. Slack (Eds.), *Managing Water Resources Past and Present* (pp. 89–116). Oxford University Press.

Kang, Y. H. (2019). A Tipping Point for Policy Transformation: Case Studies of Water Management in South Korea and Germany. In *Conference Proceedings of Adaptation Futures 2018.* Retrieved from https://doi.org/10.15641/0-7992-2543-3

Katzenstein, P. J. (1993). Coping with Terrorism: Norms and Internal Security in Germany and Japan. In J. L. Goldstein & R. Keohane (Eds.), *Ideas and Foreign Policy. Beliefs, Institutions, and Political Change* (pp. 265–295). Cornell University Press.

KFEM. (2018). *The National Government Organization Act Passed But River Management System Remains Malformed.* Retrieved January 12, 2020, from http://kfem.or.kr/?p=191244

Kim, C. S. (2010). The Dilemma of Development and Conservation. *Korean Society and Public Administration, 20*(4), 133–162.

Kim, C.-H. (2013a). Why Doesn't the Korean Water Management System Change? Based on Mahoney's Path Dependence Theory. *Korean Journal of Public Administration, 51*(4), 167–194.

Kim, Y. (2013b, December 7). Research Union Asks for Dr. Kim's Rehabilitation. *Newsis.* Retrieved from https://news.joins.com/article/12056183

Kim, C.-K., Shin, H., & Kim, H.-J. (2015). Changing Environmental Attitudes and Neo-developmentalism in South Korea. *Korean Society, 16*(1).

Koch, E. (1979, December 14). Alles geklärt? *Zeit*. Retrieved April 17, 2020, from https://www.zeit.de/1979/51/alles-geklaert

Kwon, W.-T., Baek, H.-J., Choi, K.-C., & Chung, H.-S. (2005). Study on Plans for the Establishment of Strategy on Climate Change Adaptation in Korea. *Journal of Atmosphere, 15*(4), 213–227.

Ländle unter. (1978, May 29). In *Der Spiegel* 22. Retrieved from https://www.spiegel.de/spiegel/print/d-40617885.html

LAWA. (1995). *Leitlinien für einen zukunftsweisenden Hochwasserschutz*. LAWA.

LAWA. (2008). *Strategie zur Umsetzung der Hochwasserriskomanagement-Richtlinie in Deutschland*. LAWA.

Lee, J. (1993, February 27). Environmental Protection in Europe: Rhine Is Going Back to Nature. *Kyunghyang Shinmun*

Lee, S.-Y. (2011). A Comparative Analysis of the Koreans' Perception of Environmental Conflict: Based on the 2007 and 2010 Survey Results of Korean Conflicts. *ECO, 15*(2), 81–109.

Lee, J. (2015, September 27). Tourist Attraction vs. Excessive Operation Cost…10 Years After Restoring the Cheonggye-cheon. *Yonhap News*. Retrieved January 2, 2022, from https://www.yna.co.kr/view/AKR20150921185200004

Lim, J. (2018, May 28). Integrative Water Management Without River Management. *Newsis*. Retrieved from https://news.naver.com/main/read.nhn?mode=LSD&mid=sec&sid1=102&oid=003&aid=0008620077

Lim, H.-J., & Kum, J. Y. (2018). Central-Local Government Relations and Decision-Making in Local Government: The Case of the Low Carbon Green Growth Policy. *Korean Public Administration Review, 52*(3), 185–214.

Lücken im Deich. (1954, July 28). In *Der Spiegel* 31. Retrieved from https://www.spiegel.de/spiegel/print/d-28957058.html

March, J. G., & Olsen, J. P. (1983). The New Institutionalism. Organizational Factors in Political Life. *American Political Science Review, 78*(03), 734–749. https://doi.org/10.2307/1961840

MBC. (2009, November 27). *Conversation with the President*. Retrieved from http://imnews.imbc.com/n_newssas/fullmovie/fullmovie02/2508548_16727.html

Merz, B., Elmer, F., Kunz, M., Mühr, B., Schröter, K., & Uhlemann-Elmer, S. (2014). The Extreme Flood in June 2013 in Germany. *La Houille Blanche, 1*, 5–10.

Ministry of Culture, Sports and Tourism. (2008). *Green Growth, Let's Open the Age of Mature Freedom*. Retrieved December 26, 2019, from http://www.

korea.kr/special/policyFocusView.do?newsId=148656226&pkgId=49500377&pkgSubId =&pageIndex=3

Ministry of Culture, Sports and Tourism. (2009). *Final Confirmation of the Four Rivers Project*. Retrieved December 27, 2019, from https://www.mcst.go.kr/kor/s_notice/press/pressView.jsp?pSeq=9998

Ministry of Environment. (2008). *Final Report of Research on National Climate Change Adaptation Master Plan Establishment* (In Korean: 국가 기후변화 적응 마스터플랜 수립 연구). Ministry of Environment.

Ministry of Environment. (2010). *Three Decades of Environmental Policies in Korea*. Ministry of Environment.

Ministry of Environment. (2019). *2018 Annual report of environmental statistics.* Retrieved February 1, 2020, from http://me.go.kr/home/web/index.do?menuId=123

Ministry of Land, Infrastructure and Transport. (2016). *The Long-Term Master Plan for Water Resource (2001-2020), 3rd Amendment*. MOLIT.

Noh, Hyunwoong (2012, October 7). K-Water Sues 'Critical Experts' Against the Four Rivers Project. *Hangyeoreh*. Retrieved September 2, 2019, from http://www.hani.co.kr/arti/society/society_general/541807.html

Palmer, J. (2010). Stopping the Unstoppable? A Discursive-Institutionalist Analysis of Renewable Transport Fuel Policy. *Environment and Planning C: Government and Policy, 28*(6), 992–1010. https://doi.org/10.1068/c09206j

Park, C.-K. (2009). Some Problems of the Green Growth Policy and the Project Saving the Four Major Rivers. *E&S, 83*, 117–360.

Pelling, M. (2011). *Adaptation to Climate Change. From Resilience to Transformation*. Routledge.

Polski, M. & Ostrom, E. (1999). *An Institutional Framework for Policy Analysis and Design*. Retrieved December 3, 2019, from https://pdfs.semanticscholar.org/ec83/18779f3f04c6a88cb59cdb338d4d8cde3b85.pdf

Prevent Disasters by Flood Control. (1962, August 30). In *Dong-a Ilbo*.

Professors' Organization for Movement against the Grand Korean Canal (POMAC). (2011). *A Letter of POMAC to UNEP*, Retrieved January 2, 2022, from http://hanamana.de/dul/post/410/

Radtke, I., Hustedt, T., & Klinnert, A. (2016). Inter-Ministerial Working Groups as a Panacea for Coordination Problems? *dms, 9*(1), 65–81.

Realmeter. (2010). *Support for the Four Rivers Project Is Bigger Than Opposition for the First Time*. Retrieved January 19, 2020, from http://www.realmeter.net/

Samuels, P., Klijn, F., & Dijkman, J. (2006). An Analysis of the Current Practice of Policies on River Flood Risk Management in Different Countries. *Irrigation and Drainage, 55*(S1), S141–S150. https://doi.org/10.1002/ird.257

Schmidt, V. A. (2008). Discursive Institutionalism: The Explanatory Power of Ideas and Discourse. *Annual Review of Political Science, 11*(1), 303–326. https://doi.org/10.1146/annurev.polisci.11.060606.135342

Schmidt, V. A. (2009). Comparative Institutional Analysis. In T. Landman & N. Robinson (Eds.), *Sage Handbook of Comparative Politics* (pp. 125–143). Sage.

Schmidt, V. A. (2010). Taking Ideas and Discourse Seriously: Explaining Change Through Discursive Institutionalism as the Fourth 'New Institutionalism'. *European Political Science Review, 2*(01), 1. https://doi.org/10.1017/S175577390999021X

Schmidt, V. A. (2015). Discursive Institutionalism: Understanding Policy in Context. In F. Fischer, D. Torgerson, A. Durnová, & M. Orsini (Eds.), *Handbook of Critical Policy Studies* (pp. 171–189). Edward Elgar.

Schreurs, M. (2002). *Environmental Politics in Japan, Germany, and the United States*. Cambridge University Press.

Schwarz, K. (2015). *The National Flood Protection Program: Accelerating the Implementation of Large Scale Retention Measures Within Germany*. BMU. Retrieved May 4, 2019, from https://circabc.europa.eu/webdav/CircaBC/env/wfd/Library/floods_programme_1/b_wg_f_on_floods/17th%20meeting%20-%2011-12_03_2015/Presentations/11c_National%20Flood%20Protection%20Program%20process%20to%20accelerate%20implementation%20of%20large%20scale%20retention%20measures.pdf

Steyaert, P., & Ollivier, G. (2007). The European Water Framework Directive: How Ecological Assumptions Frame Technical and Social Change. *Ecology and Society, 12*(1).

Thieken, A. H., Kienzler, S., Kreibich, H., Kuhlicke, C., Kunz, M., Mühr, B., et al. (2016). Review of the Flood Risk Management System in Germany After the Major Flood in 2013. *E&S, 21*(2). https://doi.org/10.5751/ES-08547-210251

Tu, K., Yinlong, X. & Ye, Q. (2012). *China's National Climate Change Adaptation Strategy in an International Context*. Retrieved April 15, 2020, from https://carnegieendowment.org/2012/05/18/china-s-national-climate-change-adaptation-strategy-in-international-context-event-3762

Tümmers, H. (1994). *Der Rhein. Ein europäischer Fluß und seine Geschichte*. Beck.

UBA. (2019). *Anpassung auf Bundesebene*. Retrieved December 30, 2019, from https://www.umweltbundesamt.de/themen/klima-energie/klimafolgen-anpassung/anpassung-auf-bundesebene#textpart-1

Uekötter, F. (2014). *The Greenest Nation? A New History of German Environmentalism*. The MIT Press.

Wir sollten aufwachen und überlegen. (1986, November 17). In *Der Spiegel*. Retrieved December 23, 2019, from https://www.spiegel.de/spiegel/print/d-13521501.html

Yun, S.-J., & Lee, D.-H. (2010). Agenda Setting and Frame of TV News About 4 Major Rivers Project in Korea. *ECO, 14*(1), 7–62.

Zahariadis, N. (2014). Ambiguity and Multiple Streams. In C. Weible & P. A. Sabatier (Eds.), *Theories of the Policy Process* (Vol. 2, pp. 25–58). Westview Press.

7

General Conclusion

7.1 Summary of the Argument

The purpose of this book has been to examine the factors behind the different water management approaches for adapting to climate change in South Korea and Germany by investigating their decision-making processes. Anthropogenic climate change is increasing the likelihood of extreme weather events such as typhoons, destructive storms, tropical cyclones, and hurricanes, and as a consequence, flood risk is increasing in both countries (Korea Meteorological Administration, 2017; BMU/UBA, 2018). While the two countries share similarities in social, economic, and geographic conditions, the flood risk management policies indicated in their national adaptation plans (the NCCAP and the DAS) established in the late 2000s were significantly different. Germany adopted 're-naturalisation' as their main strategy for river-based flood control, prioritising the use of retention areas and floodplains (nature-based solutions), while South Korea initiated the 'Four Rivers Project' which consisted of building dams and dredging the major rivers. The difference between the two cases is intriguing because Germany used to manage flood risk mainly through engineering-based measures similar to South Korea.

Y. h. Kang, *Climate Change Adaptation in River Management*, Palgrave Studies in Water Governance: Policy and Practice,
https://doi.org/10.1007/978-3-031-10486-2_7

Discursive institutionalism was selected as the analytical framework through which the policy process of the cases was studied. Discourse analysis and institutional analysis offer insight into two of the key elements in policy process analysis: agents and structures. Primarily, the study looked separately at discourses and institutions and their influence on the policy processes of South Korea and Germany. In keeping with grounded theory which follows an abductive research process, once the data from the two cases had been analysed, it emerged that a more powerful explanation was provided by viewing discourse and institutions simultaneously. As a result, discursive institutionalism was adopted as the analytical framework going forward.

Following Schmidt's (2010) definition of discursive institutionalism, this book first explains policy changes through ideational development and actors' interactions. Then, this study went a step beyond Schmidt's theory by attempting to explain how institutions influence policy discourses. Furthermore, this book borrowed parts of its analytical concepts from historical institutionalism in order to complement discursive institutionalism: path dependence and critical junctures created by exogenous shocks.

In terms of methodology, this study combined grounded theory methods and comparative case study. Climate change adaptation is a relatively new research field where novel theories based on empirical findings are necessary. In this context, a grounded theory approach was chosen because the main feature of grounded theory is the constant comparison of empirical findings and theories in order to establish a theory (Mills et al., 2006; Flick, 2007). In addition, the research is based on a most similar system design to find the essential factors that explain differences in flood risk management, given the similarities in the geographical and economic-political characteristics of South Korea and Germany.

Three different approaches to data collection were combined in pursuance of triangulation. Fifty-four semi-structured interviews and 13 observational activities were conducted. Relevant media reports from selected newspapers and TV programmes were also collected through keyword searching. The collected data was analysed in two qualitative ways—first, the interviews and notes from observation were coded with a data analysis software, MAXQDA. Second, the data was analysed through process tracing. Coding was a useful tool for analysing current actors' discourses and process tracing was adopted in order to understand the historical background of discourses, institutions, and policy changes.

In South Korea, the main flood management policy for climate change adaptation was the Four Rivers Project. The project was comprised of hard infrastructure measures including dams and dredging the riverbeds of the four major rivers in Korea. There are several institutional features of Korean politics forming the project's backdrop. The first feature is the developmental state institutions developed during the fast-economic growth period of the 1960s–1970s. State-led industrialisation based on the close relationship between government bureaucracy and corporations is the main characteristic of the developmental state. The institutional legacy of this developmental state has remained after the fast-growth era ended. Also, environmental policy is mostly dealt with at the national level under the strong presidency within the centralised decision-making process, and water management is not an exception.

Another institutional feature that has influenced Korean flood risk management is the water policy on the division of quality and quantity management. Although water quantity and quality issues are closely related to each other, the two issues have continued to be managed separately. This is one of the reasons why flood management based on hard infrastructure remained unchanged after shocking events in the 1990s, such as the phenol contamination incident and the waste organic solvent contamination accident in the Nakdong River. While the Ministry of Environment is tasked with water quality control, water quantity issues including dam development were handled by the Ministry of Land, Infrastructure and Transport (MOLIT). Consequently, the MOLIT, in charge of water quantity policy, led the Four Rivers Project with a heavy focus on technical and engineering measures, which worsened water quality as a result.

Besides the formal and informal institutions of the Korean political system, dominant social discourses, namely developmentalism and techno-centrism have also influenced the Four Rivers Project. The two social discourses contributed to the strong tendency of water management (programmatic discourse) and flood risk management (policy discourse) towards an engineering-development approach. Although environmental awareness has increased in Korea since the 1980s, the movements for environmental protection have not gained substantial political power. In contrast, techno-centrism and developmentalism have been more institutionalised in the political system and become resistant to change. The Four Rivers Project shows that the frames and narratives adopted from the dominant social discourses were actively used for justifying the project.

The policy process of German flood risk management also shows that discourses and institutions have been critical factors in the decision-making process. The main institutional feature of German environmental policy, including water policy, is that decisions are made at multiple levels. Due to traditional federalism, water management has been a *Länder* responsibility; the federal government has had limited power over the policy sector. As a member of the European Union, Germany also has to conform to EU regulations. Therefore, German environmental policy is referred to as being executed in one of the most complex legal systems in the world (Jänicke & Weidner, 1997).

The National Flood Protection Programme (NFPP) reflects such a complex multi-level governance system. The NFPP is the representative flood management policy of Germany adopted in 2014. Before the introduction of the NFPP, the unclear responsibilities between different government levels were pinpointed as the primary reason for several large-scale floods in the 1990s and 2000s (Thieken et al., 2016). As a consequence, the idea of solidarity among *Länder* and more coordinated governance for flood risk management has been strengthened. The NFPP was established in an attempt to enhance a more cooperative and coordinated approach through the federal government's engagement. In terms of policy goals, the NFPP follows the EU Water Framework Directive and the Floods Directive which emphasise the importance of ecosystem services in flood risk management.

Another important institutional feature of German water management is that participatory governance is regarded as a norm. In fact, corporatism based on informal cooperation between the government and interest groups dominated the decision-making process of environmental policies until the 1980s. However, as environmental movements gained support and the green party, *Die Grünen,* entered into the political sphere, the pressure grew to increase the number and the diversity of actors in the environmental policymaking process (Blühdorn et al., 1995). Consequently, participatory governance has gradually gained ground in environmental politics, and now it is regarded as a compulsory and complimentary element in the decision-making process.

The changes in dominant social discourses have played a major role in the German water management policy process. Like in South Korea, optimism for technology has influenced water management in Germany. Germans transformed their rivers by straightening and canalising for navigation, flood control, and power generation. The technology to change the course of rivers was believed to be an advancement of society. As a result, two-thirds of former floodplains have disappeared in the last 250 years (BMU & BfN, 2009). In addition, concerns over water pollution and the ecological degradation of the rivers were not frequently discussed until the 1960s; however, with the rising political power of environmental movements, environmentalism has been institutionalised in German politics. In the water management sector, integrated water resource management gradually emerged as an alternative policy discourse to the separate management of water quality and quantity. Concurrently, hydraulic engineers started to raise questions about the impact of river engineering on increasing flood risks. As a result, the importance of hydromorphology (waterbody structures) in flood protection and ecosystem conservation started to appear in the media (Bewerunge, 1988; "Hochwasser in den Flüssen", 1980).

After environmentalism became more dominant in German society, major floods provided momentum for policy change. The integrated flood risk management approach was proposed for the first time after the 1995 floods. Re-naturalisation of the rivers was promoted in Germany for aesthetic, ecological, and practical reasons. After the 2013 Elbe and Danube floods happened due to a dike breach, the NFPP was suggested. 'Giving more room to the rivers' became a popular concept not only in Germany but in other European countries including the Netherlands. Although there are still conflicts over flood risk management measures among stakeholders, mainly between farmers and ecologists, the current government policy discourse of Germany follows the ecosystem service-conservation discourse.

When the two cases are compared, it can be concluded that there are two different kinds of relationships between discourses and institutions within policy processes. On the one hand, as can be seen through developmentalism and the developmental state institution in South Korea, discourse and institutions can form a reinforcing relationship. Within the developmental state institutions, developmentalism could be easily institutionalised and become resistant. The developmental state institutions could implement pro-development policies by framing the policies with attractive storylines and metaphors which are in line with the dominant social discourse (developmentalism and techno-centrism). A strong presidency and the institutionalised top-down decision-making process of the government block the inflow of new ideas and discourses, hindering policy change.

On the other hand, discourse can also weaken existing institutions. When an alternative discourse gains substantial support, it is institutionalised within the political system leading to institutional change. Particularly, when a shocking event occurs, the inertia of sticky institutions can be undermined and a critical juncture created. In the case of Germany, environmentalism grew in prominence from the 1980s and it became a decisive factor behind the change in flood management policy. Environmental movements also influenced the expansion of participatory governance. The multi-level governance institutions allowed alternative opinions on water management policy to enter into the decision-making process in various channels. It is important to note that institutional change itself does not always lead to changes in social, programmatic, or policy discourse, even if institutions can facilitate discursive change.

The results of this study confirm the validity of discursive institutionalism as proposed by Schmidt (2008, 2010). The evidence of this study indicates that policy change and stability are the results of discourse and institutional interaction. Historical institutionalism emphasises the role of external shocks in the making of a critical juncture, but this study shows that shocks are not enough to change existing, self-reinforcing institutions. Policy change could occur when an alternative discourse gains sufficient political space AND a critical event happens. Although path dependency is the strong tendency for established institutions, shocking events together with discursive change can trigger institutional change.

7.2 Research Implications

7.2.1 Politics Matter to Adaptation

Climate change adaptation is a relatively new policy issue and how it is being addressed politically has not been thoroughly investigated. This book may help to understand why certain policies are adopted for climate change adaptation. Although climate change is an unprecedented phenomenon, many issues related to adaptation are often not necessarily new. For example, climate change is increasing the intensity and frequency of extreme weather events, and consequently, the probability of floods and drought is increasing. However, floods and drought problems have been dealt with for a long time in the water management sector. Policy reactions in response to those problems are therefore likely to occur within the context of established institutions and discourses in water management. As Eriksen et al. (2015, p. 524) state, "adaptation must be seen as part of the dynamics of societies", not as something new and separated from other socio-political issues. This study therefore analysed flood risk management policy in national adaptation plans within the historical, social and political context of water management.

There is no one-size-fits-all solution for every country or every circumstance resulting from climate change. The cases of this study support the argument that physical and geographical conditions are not the most decisive factors in flood risk management (Driessen et al., 2016). Cost-benefit analysis and technical evaluation of feasibility would be taken into account when deciding policy options, but every policy is constructed through a political process. In this study, it was revealed that social discourses, decision-making structures, informal norms, and shocking events have influenced the flood risk management policies of the two case countries. Future adaptation policy can be better designed when the policy process is better understood by stakeholders and decision-makers.

Another implication of this study for the field of adaptation is that it is hard to gain a comprehensive picture of adaptation policy by looking into adaptation plans or strategies alone. Because of the historical development of each sector, it is important to include sectoral analysis in the research. Furthermore, there are adaptation-relevant policies without using the terms 'adaptation' or 'climate change' in the policy text. With this in mind, researchers who analyse adaptation policies need to consider a broader scope of related policies beyond adaptation plans.

7.2.2 Participation Increases the Chance of Policy Change

The second implication of this study is that certain institutions are found to be more helpful to policy change than other institutions—participatory processes increase the chance of policy change, which might be an important element for future climate change adaptation. Participation of diverse stakeholders facilitates the inflow of alternative discourses to the decision-making process because they provide different kinds of knowledge which can possibly lead to "innovative solutions" (Pahl-Wostl, 2015, p. 33). Flexible and open institutions allow for learning, which is critical to increase adaptive capacity (Folke et al., 2002).

The enhanced environmental awareness in Germany after the 1980s was mentioned by several interviewees as the critical driver of changes in water management, including flood risk management. However, in cases where there was no way to institutionalise an alternative discourse, policy change was unlikely to occur after the discursive change. When *Die Grünen* entered the German parliament, traditional corporatism within German politics was opened to more diverse actors in the environmental policymaking process (Blühdorn et al., 1995). In addition, multi-level governance from the EU to the municipal level helped to increase communication channels for various actors (Wurzel, 2002; Samuels et al., 2006).

The case of Germany can be compared to South Korea which has not experienced substantial policy changes regarding flood risk management. As explained in Sect. 4.4.3, environmentalism has grown gradually in

Korea since the 1980s, but it didn't result in institutional changes like in Germany. A strong developmentalist discourse which supported the stable top-down developmental state was identified by interviewees as the reason. Although a movement against the Four Rivers Project was initiated by environmental NGOs, they could not stop the project. This is not only because their discursive power was weak, but also because the institutional setting of Korean environmental policy did not allow diverse actors to participate in the policymaking process.

In addition to facilitating policy change, participatory governance can reduce the uncertainty related to implementing a new policy because the stakeholders are informed about the policy (Newig et al., 2005). This is another important implication for climate change adaptation in general, because more new policies are expected to be planned and implemented due to continuing climate change. Political space which is open to alternative discourses and encourages active discussions provides the opportunities to examine alternative policies under different frames (Bosomworth, 2018). Inclusion of different perspectives in a decision-making process can also diversify the means and ends of solutions, which is necessary for the uncertain future with unprecedented environmental changes (Bosomworth, 2018).

7.2.3 Contribution to Water Governance Literature

This book attempted to analyse flood management policy from the perspective of adaptation. Climate change is a relatively new subject in water management and this book offers an analytical framework which can possibly be useful in future work on water management dealing with adaptation. The contextualisation of adaptation in water management through process tracing from 1950 to 2019 would be an alternative analytical framework to be used in future research.

With regard to flood risk management, this study showed that critical junctures are not always created by floods. Changes in environmental awareness and external shocks, including seemingly irrelevant events (e.g., Chernobyl disaster, ruling party change, and chemical accidents), have also influenced flood risk management. This was the case when the major water policy reform was carried out after the Sandoz accident (which occurred a few months after the Chernobyl disaster) in Germany and the new government of Moon Jae-in initiated the water institution reform in South Korea.

In addition, the consequence of certain policies can open a window of opportunity for policy change, as can be seen in the case of South Korea. After the completion of the Four Rivers Project, algal bloom in the rivers gained public attention. Although the problem existed even before the Four Rivers Project, the project triggered public concerns about the causes of the algal blooms and drinking water quality and this in turn made them question river management (Kim, 2012). The argument for more integrated river management emerged as an alternative policy discourse. Such change in policy discourse was recognised by politicians during the 2017 presidential election. After the opposition party won the election, water management policy changed in a relatively short time. First, an institutional reform in the water sector was carried out: The Ministry of Environment took over water management tasks from the MOLIT, although the tasks related to river management remained with them ("Summary" 2018). Since this formal institutional change, the government of Korea has also discussed dismantling some of the dams built during the Four Rivers Projects. The project can therefore be regarded as a window of opportunity for policy change.

Lastly, this book contributes to systematise the discursive analysis of water management by proposing a categorisation and a triangular discursive space where the programmatic discourses are located (Sect. 3.2.3). With the primary emphasis on flood control measures, this book suggested dividing water management discourses into four categories: engineering-development, participation-equity, conservation-ecosystem service, and integrated water resource management (IWRM). Those four discourses form a triangular space of water management discourse, which can be utilised for examining various water management cases.

7.3 A Future Research Agenda Building on the Findings of This Research

This study's ability to generalise is limited by the number of cases. Because both Germany and South Korea are industrialised countries, the findings could be useful for other industrialised countries. Findings should, however, be applied with careful attention to the uniqueness of the two cases. For example, both countries have had strong tendencies to be optimistic about the role technology can play in water management. This powerful discourse has a historical background: the development imperative of economic and ideological competition after the Second World War. The resultant environmental problems, including damage to ecosystems, water pollution, and loss of communities, were treated by policymakers as the side-effects of essential activities needed for economic growth.

This study was enriched by interviews, observation, and media analysis. Discourse analysis was undertaken with qualitative data, mainly collected through semi-structured interviews and media reports. As it was not possible to reach all relevant actors of flood risk management policies, some crucial actors' statements had to be collected through media and online search (e.g., Lee Myung-bak, the former president of South Korea). Nevertheless, this study offers some insight into how discourse and institutions interact in the process of adaptation policy process. In the future, public opinion surveys and interviews with a broader extent of stakeholders could lead to yet deeper insights into the dynamics of the adaptation policy process.

This study focused on water management, especially on flood risk management, just one of the many policy sectors indicated in national climate change adaptation plans. Future research could examine policy responses (change or stasis) for adapting to climate change in other areas. Also, as the scope of this analysis was limited to the policy planning process and did not cover the implementation of the plans, future research could extend its scope to the implementation and evaluation processes. While some interviewees did point out that adaptation plans did not always correspond with policy implementation in practice, the discordance was not dealt with in this study.

The findings of this study suggest that the policy process is influenced by the interactions of discourse and institutions, which is in accordance with discursive institutionalism. The findings from South Korea and Germany led to the discourse-institution policy process framework suggested in Fig. 6.5. Future research might explore other countries' adaptation policies or water management policies with the policy process framework, which can be useful for investigating the reasons behind policy change and stasis.

Several questions were triggered by this research. How much has climate change affected the discourses and institutions related to flood risk management? Can the case of flood risk management be applied to other sectors of adaptation? Can the importance of external shocks, changes in discourses, and institutional settings in policy processes be quantified and compared with other cases? What other factors beyond discourses and institutions influence policy change and stasis? Further studies could address these questions.

Adaptation to climate change is not a one-time goal. As climate change is very likely to continue in the future, human society should keep searching for ways to adapt. Therefore, adaptation is a continuous process. This implies that the effort to conduct research on adaptation should also be maintained. Improving public policy for better adaptation is heavily dependent on persistent efforts to uncover the logic of policy processes. Hopefully, the empirical and theoretical implications of this study may contribute to the continuing search for sustainable adaptation for all.

References

Bewerunge, L. (1988, March 31). Die Fluten fließen zu schnell. *Frankfurter Allgemeine Zeitung*, p. 10.

Blühdorn, I., Krause, F., & Scharf, T. (1995). Introduction: Getting Ready for the Third Phase. In I. Blühdorn, F. Krause, & T. Scharf (Eds.), *The Green Agenda. Environmental Politics and Policy in Germany* (pp. 11–24). Keele University Press.

BMU & BfN. (2009). *Auenzustandbericht, Flussauen in Deutschland*. BfN.

BMU/UBA. (2018). *Water Resource Management in Germany. Fundamentals, Pressures, Measures*. UBA.
Bosomworth, K. (2018). A Discursive–Institutional Perspective on Transformative Governance: A Case from a Fire Management Policy Sector. *Environmental Policy and Governance, 28*(6), 415–425. https://doi.org/10.1002/eet.1806
Driessen, P. P. J., Hegger, D. L. T., Bakker, M. H. N., van Rijswick, H. F. M. W., & Kundzewicz, Z. W. (2016). Toward More Resilient Flood Risk Governance. *E&S, 21*(4). https://doi.org/10.5751/ES-08921-210453
Eriksen, S., Nightingale, A. J., & Eakin, H. (2015). Reframing Adaptation. The Political Nature of Climate Change Adaptation. *Global Environmental Change, 35*, 523–533. https://doi.org/10.1016/j.gloenvcha.2015.09.014
Flick, U. (2007). *Managing Quality in Quantitative Research*. Sage.
Folke, C., Carpenter, S., Elmqvist, T., Gunderson, L., Holling, C. S., & Walker, B. (2002). Resilience and Sustainable Development: Building Adaptive Capacity in a World of Transformations. *Ambio, 31*(5), 437–440.
Hochwasser in den Flüssen zu einem Gutteil hausgemacht. (1980, November 2). In *Frankfurter Allgemeine Zeitung*, p. 9.
Jänicke, M., & Weidner, H. (1997). Germany. In M. Jänicke, H. Weidner, & H. Jörgens (Eds.), *National Environmental Policies. A Comparative Study of Capacity-Building: With a Data Appendix, International Profiles of Change Since 1970* (pp. 133–156). Springer.
Kim, H. (2012, August 8). The Four Rivers Became "Green Tea Latte Disaster". *Nocut News*. Retrieved from https://news.naver.com/main/read.nhn?mode=LSD&mid=sec&sid1=100&oid=079&aid=0002383498
Korea Meteorological Administration. (2017). *Climate Change Projection Report in the Korean Peninsula for the New Climate Regime*. KMA
Mills, J., Bonner, A., & Francis, K. (2006). The Development of Constructivist Grounded Theory. *International Journal of Qualitative Methods, 5*(1), 25–35.
Newig, J., Pahl-Wostl, C., & Sigel, K. (2005). The Role of Public Participation in Managing Uncertainty in the Implementation of the Water Framework Directive. *European Environment, 15*(6), 333–343. https://doi.org/10.1002/eet.398
Pahl-Wostl, C. (2015). *Water Governance in the Face of Global Change*. Springer.
Samuels, P., Klijn, F., & Dijkman, J. (2006). An Analysis of the Current Practice of Policies on River Flood Risk Management in Different Countries. *Irrigation and Drainage, 55*(S1), S141–S150. https://doi.org/10.1002/ird.257

Schmidt, V. A. (2008). Discursive Institutionalism: The Explanatory Power of Ideas and Discourse. *Annual Review of Political Science, 11*(1), 303–326. https://doi.org/10.1146/annurev.polisci.11.060606.135342

Schmidt, V. A. (2010). Taking Ideas and Discourse Seriously: Explaining Change Through Discursive Institutionalism as the Fourth 'New Institutionalism'. *European Political Science Review, 2*(01), 1. https://doi.org/10.1017/S175577390999021X

Thieken, A. H., Kienzler, S., Kreibich, H., Kuhlicke, C., Kunz, M., Mühr, B., et al. (2016). Review of the Flood Risk Management System in Germany After the Major Flood in 2013. *E&S, 21*(2). https://doi.org/10.5751/ES-08547-210251

Wurzel, R. (2002). *Environmental Policy-Making in Britain, Germany and the European Union. The Europeanisation of Air and Water Pollution Control.* Manchester University Press.

Appendix 1. List of Interviewees

South Korea

Anonymous 1, public officer, Ministry of Environment, Sejong, 9 March 2017.

Anonymous 2, public officer, Ministry of Environment, Sejong, 9 March 2017.

Anonymous, public officer, Ministry of Agriculture, Food and Rural Affairs, Sejong, 9 March 2017.

Anonymous 1, senior researcher, Gyeonggi Research Institute, Suwon, 13 March 2017.

Anonymous 2, senior researcher, Gyeonggi Research Institute, Suwon, 13 March 2017.

Anonymous, senior researcher, Korea Institute of Civil Engineering and Building Technology, Seoul, 14 April 2017.

Anonymous, senior researcher, Korea Environment Institute, Sejong, 2 March 2017.

Anonymous, senior researcher, Korea Institute for Health and Social Affairs, Sejong, 15 March 2017.

Y. h. Kang, *Climate Change Adaptation in River Management*, Palgrave Studies in Water Governance: Policy and Practice,
https://doi.org/10.1007/978-3-031-10486-2

Anonymous, former researcher at Korea Institute of Civil Engineering and Building Technology, Goyang, 14 March 2017.

Anonymous, public officer, Ministry of Public Safety and Security, Sejong, 9 March 2017.

Anonymous, public officer, Ministry of Land, Infrastructure and Transport, Seoul, 6 March 2017.

Anonymous, public officer, Ministry of Land, Infrastructure and Transport, Seoul, 6 March 2017.

Anonymous, public officer, Rural Development Administration, Suwon, 15 March 2017.

Anonymous, public officer, Ministry of Environment, Sejong, 2 March 2017.

Anonymous, public officer, K-water, Sejong, 15 March 2017.

Ahn Byungok, founder, Climate Action, Seoul, 28 February 2017.

Ahn Sook-Hee, Korea Federation for Environmental Movements (KFEM), Seoul, 16 March 2017.

Hong Jong Ho, professor, Seoul National University, Seoul, 21 March 2017.

Jo Hongseop, environmental journalist, Seoul, 10 April 2017.

Jung Soo Gun, KFEM Daegu, Daegu, 10 March 2017.

Kang Yanggu, environmental Journalist, Seoul, 14 April 2017.

Kim Young-Oh, professor, Department of Civil and Environmental Engineering, Seoul National University, Seoul, 20 March 2017.

Kim Kibum, environmental journalist, Kyunghyang Shinmun, 7 March 2017.

Kim So Hee, Climate Change Center, Seoul, 16 March 2017.

Lee Young Kee, lawyer, Seoul, 6 March 2017.

Lee Jung-Hwan, National Assembly Forum on Climate Change, Seoul, 3 March 2017.

Park Seok Soon, professor, Ewha Womans University, Seoul, 7 March 2017.

Shin Jenny, KFEM, Seoul, 16 March 2017.

Jill Leslie Tao, professor, Department of Public Administration, Incheon National University, Seoul, 27 February 2017.

Yun Sun-Jin, professor, Seoul National University, Seoul, 6 March 2017.

Yum Hyung-cheol, Secretary General, KFEM, Seoul, 16 March 2017.

Germany

Anonymous, geo-ecologist, a state environment agency, 11 July 2017.

Anonymous, a state environment agency, 11 July 2017.

Anonymous, senior researcher, Humboldt University of Berlin, Berlin, 1 November 2017.

Corinna Baumgarten, UBA (German Environment Agency), Dessau, 20 July 2017.

Britta-Antje Behm, Climate Protection and the Energy Sector International Agreements Rhine and Danube, European Water Framework Directive, Support Schemes, Ministry of the Environment of Baden-Württemberg, Stuttgart, 20 November 2017.

Michael Bender, Grüne Liga, Berlin, 16 November 2017.

Achim Daschkeit, UBA KomPass—Kompetenzzentrum Klimafolgen und Anpassung, telephone interview, 22 November 2017.

Jörg Freyhof, Leibniz Institute of Freshwater Ecology and Inland Fisheries (IGB), Berlin, 2 March 2018.

Hans Helmut Bernhart, professor, Karlsruher Institute for Technology Institut für Wasser und Gewässerentwicklung, Fachbereich Wasserwirtschaft und Kulturtechnik, Karlsruhe, 18 July 2017.

Martin Grambow, Water Management and Soil Protection Head, Bavarian State Ministry of the Environment and Consumer Protection, 8 November 2017.

Björn M. Grün, Emschergenossenschaft/Lippeverband (EGLV) Management, Essen, 7 November 2017.

Jesko Hirschfeld, professor, Institute for Ecological Economy Research, Berlin, 16 October 2017.

Edith Hödl, International Commission for the Protection of the Danube River, Technical Expert in River Basin Management, telephone interview, 16 April 2018.

Ulrike Hursie, Head of the Secretariat, Elbe River Basin Community Liaison Office (FGG Elbe), Magdeburg, 23 November 2017.

Stephanie Korte, Climate Protection and the Energy Sector International Agreements Rhine and Danube, European Water Framework Directive, Support Schemes, Ministry of the Environment of Baden-Württemberg, Stuttgart, 20 November 2017.

Hyejee Lim, independent activist, Munich, 17 July 2017.

Ralf Ludwig, professor, Department of Geography of Ludwig-Maximilians-Universitaet Muenchen, Munich, 21 June 2017.

Nicole Mahlkow, Environmental Policy Research Center, Free University of Berlin, Berlin, 12 July 2017.

Sascha Maier, BUND-Brandenburg, Berlin, 24 January 2018.

David Moryc, Senior Director, American Rivers, Berlin, 21 January 2018.

Sandra Naumann, Elbe River Basin Community Liaison Office (FGG Elbe), Magdeburg, 23 November 2017.

Ekkehard Pfeiffer, Head of Coordination Unit for Corporate Tasks, Emschergenossenschaft/Lippeverband (EGLV), Essen, 7 November 2017.

Anne Schulte-Wülwer-Keidig, Head of the Secretariat, International Commission for the Protection of the Rhine (ICPR) Secretariat, Koblenz, 27 October 2017.

Lemke Steffi, Parliament member, German federal parliament Bündnis 90/Die Grünen, email interview, 6 July 2018.

Netherlands

Jos van Alphen, Deltacommissaris, telephone interview, 24 January 2020.

Nathalie Asselman, specialist, Deltares, Delft, 8 November 2018.

Dave Huitema, professor, Vrije Universiteit Amsterdam, Utrecht, 7 November 2018.

Maria Kaufmann, senior researcher, Radboud University, video interview, 25 October 2018.

Erik Mosselman, specialist, Deltares, Delft, 8 November 2018.

Maarten Ouboter, Waternet (Amsterdam Regional Water Authority), Amsterdam, 29 November 2019.

Gerhard van den Top, Chair, Waternet, Amsterdam, 17 December 2019.

Appendix 2. Conferences and Public Events Attended for the Research

11th International Conference of the European Society for Ecological Economics, Leeds, UK, 30 June–3 July 2015.

Berlin Conference 2016: Transformative Global Climate Governance après Paris, Berlin, Germany, 23–24 May 2016.

International Political Science Association: 24th World Congress of Political Science, Poznan, Poland, 23–28 July 2016.

Water Reform Forum: Workshop on the ways of reforming Korea's water management, Seoul, South Korea, 3 March 2017.

KFEM: Workshop on the Four Rivers Project: task and direction suggestions for the next president, Seoul, South Korea, 17 March 2017.

Ludwig Maximilian University of Munich: 1st ClimEx Symposium-Climate Change Impacts on Extreme Events: Risks and Perspectives for Water Management, Munich, Germany, 20–21 June 2017.

3rd European Climate Change Adaptation Conference, Glasgow, UK, 5–9 June 2017.

Resilience 2017, Stockholm, Sweden, 20–23 August 2017.

Impacts World 2017, Potsdam, Germany, 11–13 October 2017.

Grüne Liga: Flussfilmfest Berlin 2018, Berlin, Germany, 19–21 January 2018.

Y. h. Kang, *Climate Change Adaptation in River Management*, Palgrave Studies in Water Governance: Policy and Practice,
https://doi.org/10.1007/978-3-031-10486-2

Leibniz-Institut für Raumbezogene Sozialforschung: Kulturelle Grenzen im Fluss? 20 Jahre nach der Oderflut, Berlin, Germany, 3 November 2017.

5th International Climate Change Adaptation Conference, Cape Town, South Africa, 18–21 June 2018.

2018 Utrecht Conference on Earth System Governance, Utrecht, Netherlands, 5–8 November 2018.

Index[1]

[1] Note: Page numbers followed by 'n' refer to notes.

Y. h. Kang, *Climate Change Adaptation in River Management*, Palgrave Studies in Water Governance: Policy and Practice,
https://doi.org/10.1007/978-3-031-10486-2

The manufacturer's authorised representative in the EU is Springer Nature Customer Service Centre GmbH, Europaplatz 3, 69115 Heidelberg, Germany. If you have any concerns regarding our products, please contact ProductSafety@springernature.com

Printed and bound by CPI Group (UK) Ltd, Croydon, CR0 4YY
30/09/2026
02225243-0006